An Introduction to Immunotoxicology

An Introduction to Immunotoxicology

JACQUES DESCOTES

Lyon Poison Centre and Department of Pharmacology, Medical Toxicology and Environmental Medicine, INSERM U999, Lyon-RTH Laennec Faculty of Medicine, Lyon, France

UK   Taylor & Francis Ltd, 1 Gunpowder Square, London, EC4A 3DF
USA   Taylor & Francis Inc., 325 Chestnut Street, Philadelphia, PA 19106

British Library Cataloguing in Publication Data

A catalogue record for this book is available from the British Library.

ISBN 0–7484–0306–x (cased)
ISBN 0–7484–0307–8 (paperback)

Library of Congress Cataloguing-in-Publication Data are available

Cover design by Hybert Design and Type

Typeset in 10/12pt Times by Graphicraft Limited, Hong Kong

Printed and bound by T.J. International Ltd, Padstow, UK

Cover printed by Flexiprint, Lancing, West Sussex

This book is dedicated to
Christiane, Jérôme and Aurélie

Contents

Preface

Immunotoxicology is a new area of toxicology. The introduction of potent immunosuppressive drugs into the clinical setting in the 1960s was the first attempt in the
pharmacological manipulation of the immune system. Following the introduction of these
drugs, enormous progress could be achieved in the field of organ transplantation, but
severe adverse consequences to the health of treated patients were also soon evidenced.
Clinicians and toxicologists became aware that immunosuppression can result in various
infectious complications and more frequent lymphomas as well as other virus-induced
neoplasias. Concern arose that chemical exposure, either at the workplace or via environmental pollution, could induce similar adverse health consequences.

In the 1970s and 1980s, extensive research efforts were devoted to identifying the
possible immunosuppressive effects of xenobiotics, such as medicinal products, pesticides,
industrial chemicals, or food additives. That many xenobiotics can impair the immune
system's responsiveness was increasingly recognised. At the same time, improved methods and strategies for non-clinical immunotoxicity evaluation were designed. Then,
immunotoxicologists focused on other critical aspects of immunotoxicity, such as autoimmunity and hypersensitivity. Risk assessment, immunotoxicity in wildlife species and
human (epidemiological) immunotoxicology are the most recent areas of interest and
research. During the first 20 years of its existence, immunotoxicology grew to become a
fully acknowledged subdiscipline of toxicology. Nowadays, every toxicology meeting
includes sessions on immunotoxicology and the number of published papers is steadily
increasing.

However, because immunotoxicology is still a young discipline, toxicologists from
other areas of expertise often consider immunotoxicology to be highly specialised, complex and sometimes tricky. This book is an attempt to show that immunotoxicology is
in no way different from other areas of toxicology, and therefore that it is amenable
to any toxicologist. This book is primarily targeted to students and toxicologists with no
prior experience or expertise in immunotoxicology, who wish to acquire the basic knowledge for a better understanding of current immunotoxicological data, issues, methods
and needs.

This book is divided into three parts. The first part deals with the adverse health
consequences of immunotoxicity. Even though much remains to be done to gain a

comprehensive knowledge of immune-mediated clinical toxicities, a reasonable picture can be drawn in quite a few instances. The second part deals with methods: because regulatory aspects are so critical in the safety evaluation of xenobiotics, this is the first topic to be addressed. Methods to predict possible immunotoxic effects are then overviewed. The third part deals with newer aspects of immunotoxicology. At the end of each chapter, a list of selected references is provided, including major general references to suggest further reading, and illustrative references of issues raised in the text.

It is hoped that this introduction to immunotoxicology will encourage more toxicologists to become interested in this new and exciting field.

Jacques Descotes
Lyon, Saint Jean d'Avelanne

PART I

Immunotoxic Effects and their Clinical Consequences

1

Definition and History of Immunotoxicology

Definition

In the broadest sense, immunotoxicology can be defined as the science of poisons for the immune system. In laboratory animals, various toxic exposures have indeed been found to induce changes of varying magnitude – essentially decreases – in the weight of lymphoid organs (mainly the thymus, spleen, and lymph nodes), sometimes associated with corresponding disturbances of the normal histological architecture, and also to induce varied functional alterations of the immune response as well as impaired resistance to experimental infections and/or tumours.

During the International Seminar on the Immunological System as a Target for Toxic Damage held in Luxembourg on 6–9 November 1984, a seminar which is widely held as a hallmark in the short history of the discipline, the following definition was proposed: '*Immunotoxicology is the discipline of toxicology which studies the interactions of xenobiotics with the immune system resulting in adverse effects*' (Berlin *et al.*, 1987). This definition, which can be considered as the starting point to past and present efforts paid to developing immunotoxicity evaluation and immunotoxicity risk assessment, is still valid today.

The following aspects of immunotoxicology are essential:

- Immunotoxicology is genuinely an area of toxicology and the early discussion whether immunotoxicology is more an *immuno*toxicological than an immuno*toxicological* discipline, sounds totally obsolete today.

- Immunotoxicology is not merely devoted to the identification of changes in the immune system and immune responses; mechanisms of immunotoxic effects should also be considered.

- The ultimate focus is on the adverse consequences of immune changes to the health of exposed living beings.

Therefore, immunotoxicology is a multidisciplinary area. Obviously, the first line of workers in this field comprises of toxicologists and also immunologists, but other key specialists include epidemiologists, clinicians, occupational and environmental physicians, allergologists, veterinarians, molecular and cellular biologists, and statisticians.

History

Immunotoxicology is a young discipline (Abrutyn, 1979; Burleson and Dean, 1995). The first symposium ever devoted to immunotoxicology was organised in Lyon (France) on 24 October 1974 (Groupement Lyonnais de Lutte contre les Intoxications, 1975). However, the first systematic attempt to cover immunotoxicity evaluation was held in 1979 in the USA under the auspices of the New York Academy of Sciences, and this was followed by a special issue of the Journal *Drug and Chemical Toxicology* (Special Issue, 1979), which can be considered as one of the foundation texts of the discipline. The birth of immunotoxicology was formally announced by Davies in 1983.

Despite a life-span which is hardly more than 20 years, the history of immunotoxicology can be divided into several phases reflecting the evolution of the field to the present situation.

The prehistorical phase of immunotoxicology

Very early works dating back to the first days of this century or even the last days of the nineteenth century, can be found in the scientific literature. Examples of such papers dealing with the adverse influences of toxicants on the immune system include the negative role of ethanol upon the resistance of rabbits towards experimental streptococcal infections (Abbott, 1896), the impaired resistance of laboratory animals to experimental resistance following exposure to ether, chloroform or chloral hydrate (Snel, 1903), and the depressive effects of ether on phagocytosis (Graham, 1911), or those of sodium salicylate on antibody production (Swift, 1922).

In fact, the first real impetus to the study of the adverse effects of medicinal drugs and chemicals on the immune system was the introduction of potent immunosuppressive drugs into the clinical setting. Although the term 'immunotoxicology' was not used at that time, a number of papers in the 1960s described the adverse effects of immunosuppressive agents, particularly infectious complications in kidney transplant patients and in cancer patients on chemotherapy (Meyler, 1966). Therefore, immunotoxicology was initially considered exclusively from the viewpoint of unwanted immunosuppression, an approach which has long been held by many workers, particularly in the USA. Unsurprisingly, the first major literature review on immunotoxicity focused on immunosuppression as related to toxicology (Vos, 1977). Ongoing discussions dealing with possible or proposed guidelines for the non-clinical evaluation of immunotoxicity today also largely deal with immunosuppression.

In the 1970s, many papers were published essentially describing immunosuppressive effects from immunologically-oriented studies taking little, if any, care of the magnitude, route and duration of exposure (for review see Descotes, 1986). The relevance of these findings is obviously limited so that extreme caution should be exercised when using these early data for immunotoxicity risk assessment.

The American phase of immunotoxicology

The late 1970s and early 1980s witnessed a critical turning point in the development of immunotoxicology with the introduction of basic toxicological concepts, such as the selection of relevant doses, the route and duration of exposure, and animal species, into non-clinical immunotoxicity evaluation studies. The concept of tiered protocols, which is nowadays widely accepted in the non-clinical evaluation of immunotoxicity, was introduced at that time by American authors (Dean *et al.*, 1979).

Importantly, the second phase of immunotoxicology proved essential to rationalise the non-clinical immunotoxicity evaluation of xenobiotics. A number of model compounds, such as cyclophosphamide, benzidine, diethylstilboestrol, organotin derivatives, hexachlorobenzene, selected polycyclic aromatic hydrocarbons (e.g. dimethylbenz[a]nthracene) and pesticides (e.g. chlordane) were investigated, in particular within the context of an interlaboratory standardisation study in B6C3F1 mice under the auspices of the US National Toxicology Program (Luster *et al.*, 1988).

Several important symposia devoted to immunotoxicology, which proved very helpful to the development of the discipline, were held in Research Triangle Park (Special Issue, 1982) and the University of Surrey (Gibson *et al.*, 1982) in 1982, and as already indicated in Luxembourg in 1984 (Berlin *et al.*, 1987). However, despite significant methodological advances, immunotoxicology was still only considered from the viewpoint of immunosuppression, and new inputs were deemed necessary to expand the scope of immunotoxicology.

The European phase of immunotoxicology

In the late 1980s, the scope of immunotoxicology indeed expanded to include other aspects essentially under the impetus of European authors, as exemplified by the International Symposium on the Immunotoxicity of Metals and Immunotoxicology held in Hanover in 1989 (Dayan *et al.*, 1990).

Autoimmunity

Autoimmunity was the first new area to be addressed and more attention was indeed paid to drug-induced autoimmune reactions, such as the lupus syndrome, autoimmune haemolytic anaemia, or myasthenia (Kammüller *et al.*, 1989). Autoimmune reactions became a matter of concern for immunotoxicologists after several medicinal products were withdrawn from the market because of severe autoimmune reactions, such as the oculo-cutaneous syndrome induced by the beta-blocker practolol (Behan *et al.*, 1976), Guillain-Barré syndrome to the antidepressant zimeldine (Nilsson, 1983), or autoimmune haemolytic anaemia to the antidepressant nomifensine (Salama and Mueller-Eckhardt, 1986), to give a few illustrative examples only.

The assumption based on clinical symptoms and experimental data that the Toxic Oil Syndrome which affected more than 20 000 persons in Spain in 1981 (with 1518 registered deaths), could have an autoimmune origin (Gómez de la Cámara *et al.*, 1997), and more recently the eosinophilia–myalgia outbreak related to L-tryptophan intake with over 1500 victims in the USA alone (Belongia *et al.*, 1992), demonstrated that autoimmune reactions to xenobiotics, although relatively, if not very, uncommon, can nevertheless result in major health problems with significant morbidity and mortality.

Hypersensitivity reactions

Surprisingly, significant efforts began to be paid only recently to the prediction of hypersensitivity or immuno-allergic reactions, possibly because these reactions have long been considered unpredictable in animals, even though well-standardised animal models had been in use for years, for example to identify contact sensitisers (Kimber and Maurer, 1996).

Nevertheless, hypersensitivity reactions have been one of the leading causes of medicinal product withdrawal from the market during the past two decades. Examples include the intravenous general anaesthetic alfadione which induced anaphylactoid shocks in up to 800 anaesthetised patients (Radford *et al.*, 1982), the non-steroidal anti-inflammatory drugs isoxicam which caused severe toxidermias, such as the Steven–Johnson syndrome and toxic epidermal necrolysis (Fléchet *et al.*, 1985), and zomepirac, the clinical development of which was stopped because of anaphylactic shocks observed in patients from phase III trials (Sandler, 1985), and finally the minor analgesic glafenin which produced both anaphylactic and essentially non-anaphylactic (or 'pseudo-allergic') shocks (Stricker *et al.*, 1990).

The next phase: from today to tomorrow

Recently, emphasis was given on several new aspects of immunotoxicology, which are likely to account for dramatic changes in the next few years in the area of immunotoxicity evaluation.

Human immunotoxicology

Human or clinical immunotoxicology is an area of growing interest (Burrel *et al.*, 1992; Newcombe *et al.*, 1992). Several panels of experts were convened to discuss and select those markers more likely to be useful to identify immunotoxic effects in groups of the population occupationally and/or environmentally exposed to chemicals, but also to follow up such groups of exposed individuals (National Research Council, 1992; Straight *et al.*, 1994). Available biomarkers of immunotoxicity are still of limited value (Descotes *et al.*, 1996), but advances can be expected from the introduction of more sensitive and specific endpoints, especially those derived from molecular biology techniques, even though extensive efforts should certainly be paid to the standardisation, validation and cost-effectiveness evaluation of candidate assays.

Epidemiology is increasingly considered an important new area of immunotoxicology. Until recently, the main focus of immunotoxicologists was on immunosuppression and as already mentioned the health consequences of marked immunosuppression became evidenced soon after the introduction of immunosuppressive drugs into the clinical setting. More limited knowledge is available regarding the consequences of mild to moderate immunosuppression ('immunodepression') and epidemiological studies are expected to be helpful to show whether and if so, to what extent, immunodepression related to chemical exposure is associated with adverse health consequences (Van Loveren *et al.*, 1997).

Immunotoxicity risk assessment

Immunotoxicity risk assessment is another recent breakthrough in the field of immunotoxicology (Selgrade *et al.*, 1995). As in other areas of toxicology, the focus of toxicologists had been for years on the identification of hazards (i.e. toxicity) so that the risks involved under actual conditions of human exposure were not adequately addressed even though safety (or uncertainty) factors were used in the determination of acceptable dose or exposure levels. The development of methodologies for assessing and quantifying toxic risks is an important new area of immunotoxicology.

Other aspects

Several other issues have also recently begun to be addressed. The search for alternative methods is a priority for toxicologists and immunotoxicologists as well, but the majority of available *in vitro* models have not proved so far to be helpful additions to current immunotoxicity evaluation as in many other areas of toxicology (Sundwall *et al.*, 1994).

Because chemical damages to the environment are a matter of growing concern, the study of immunotoxic effects in wildlife species is expanding (Briggs *et al.*, 1996). Experience gained in wild species, such as seals from the North Sea, is certainly important to substantiate the adverse effects of immunotoxic exposure on the health status of mammalian species including man.

Regulatory aspects are critical in toxicology and a further significant step is likely to be the release of guidelines for the immunotoxicity evaluation of drugs and chemicals. This is expected to be a major new impetus for further immunotoxicological investigations both in industry and in academia (Descotes, 1998). However, it should be stressed that past and ongoing discussions and validation efforts have not been successful in delineating realistic, cost-effective and adequate regulatory requirements which have been awaited for the past ten years at least.

The field of immunotoxicology has thus been markedly expanding in the past years to encompass all aspects of the possible toxic influences on the immune system of living beings. Immunotoxicology is now widely recognised as a mature area of toxicology. A number of toxicology societies, such as the (US) Society of Toxicology, the British Toxicology Society and Eurotox (the European Society of Toxicology), established speciality sections devoted to immunotoxicology. Several journals, such as Academic Press's *International Journal of Immunopharmacology*, Marcel Dekker's *Immunopharmacology and Immunotoxicology*, and Elsevier's *Toxicology* have included immunotoxicology within their editorial scope.

Scope of immunotoxicology

The scope of immunotoxicology can be defined as the sum of the following three distinctive areas: direct immunotoxicity, hypersensitivity and autoimmunity. Typically, immunotoxicology used to be divided into immunosuppression and immunostimulation, the latter resulting in hypersensitivity and autoimmunity. Because immunostimulation can cause adverse effects with no pathophysiological relations to hypersensitivity and autoimmunity, and as similar fundamental mechanisms are presumably involved in several aspects of hypersensitivity and autoimmunity, it seems more adequate to differentiate direct immunotoxicity, hypersensitivity and autoimmunity into three distinctive areas deserving appropriately targeted modalities of non-clinical as well as clinical evaluation.

Realistically, immunotoxic effects can indeed only be reliably identified, their mechanisms reasonably understood and their onset properly predicted, if they are investigated from the separate perspectives of these three distinctive areas.

Direct immunotoxicity

As with any other physiological system or function, the immune system or immune responses can be impaired or stimulated. Direct immunotoxicity refers to quantitative changes of an otherwise qualitatively normal immune response. Changes include decreases (namely immunosuppression) or increases (immunostimulation) in the immune response. Depending on whether the immune response is totally abrogated or only partly impaired, the terms 'immunosuppression' and 'immunodepression' can be proposed to be used respectively. As the frequency and severity of resulting health consequences are somewhat different, the distinction between immunosuppression and immunodepression should be made and carefully taken into account when designing immunotoxicity evaluation protocols. Alternatively, the terms 'immunosuppression' and 'overimmunosuppression', instead of immunodepression and immunosuppression, have been sometimes used. 'Immunomodulation' was also often used instead of immunostimulation. Experimental findings that immunosuppressive drugs, such as cyclophosphamide and cyclosporine, can exert immunoenhancing effects depending on the dose regimen in respect of the time of antigen injection, indeed led to misleading expectations regarding therapeutic relevance of these findings in humans. In addition, the introduction of recombinant cytokines into the clinical setting clearly showed that immune responses as any other physiological responses can be enhanced or stimulated by pharmacological manipulation.

Hypersensitivity

Hypersensitivity reactions to drugs and industrial chemicals are relatively common in man, and often considered to be so increasingly frequent as to become a major health problem in relation to occupational or environmental chemical exposures (Salvaggio, 1990; Vos *et al.*, 1995). Fortunately, they are seldom severe or even life-threatening, and the majority are mild to moderate, and self-limiting. The clinical manifestations noted in patients developing hypersensitivity reactions are manifold, with the skin and the lung being the most commonly affected target organs, but the liver and the kidney can also be involved.

A few mechanisms, such as T-lymphocyte mediated contact hypersensitivity or IgE-mediated anaphylaxis, have been identified to be involved, and the prediction of hypersensitivity reactions involving such mechanisms is indeed possible to some extent in laboratory animals. However, the mechanism of most hypersensitivity reactions remains ill-explained, if not totally unknown, so that prediction in animals or diagnosis in humans is very difficult or even impossible in many instances. Renewed research efforts should be (and actually are beginning to be) put into the investigation of the many possible mechanisms involved. As hypersensitivity reactions are not immune-mediated in all instances, investigations should also focus on non-immune-mediated ('pseudo-allergic') mechanisms.

Autoimmunity

Autoimmune reactions to drugs and chemicals seem to be rare events (Vial *et al.*, 1994) despite undocumented claims from a number of authors that medicinal products and chemicals are frequent but ignored causes of autoimmune diseases, in particular systemic autoimmune diseases, such as lupus erythematosus.

The mechanisms involved are largely unknown as are the fundamental mechanisms of autoimmunity in general. Major advances in our understanding of organ-specific auto-immune reactions seem unlikely in the next few years so that predicting these reactions related to medicinal drug therapy and chemical exposure can be expected to remain beyond reach for many years. Some progress was recently achieved regarding the prediction of systemic autoimmune reactions.

References

ABBOTT, A.C. (1896) Acute alcoholism on the vital resistance of rabbits to infection. *J. Exp. Med.*, **1**, 447–453.

ABRUTYN, E. (1979) The infancy of immunotoxicology. *Ann. Intern. Med.*, **90**, 118–119.

BEHAN, P.O., BEHAN, W.M.H., ZACHARIAS, F.J. and NICHOLLS, J.T. (1976) Immunological abnormalities in patients who had the oculomucocutaneous syndrome associated with practolol therapy. *Lancet*, **ii**, 984–987.

BELONGIA, E.A., MAYENO, A.N. and OSTERHOLM, M.T. (1992) The eosinophilia-myalgia syndrome and tryptophan. *Annu. Rev. Med.*, **12**, 235–256.

BERLIN, A., DEAN, J.H., DRAPER, M.H., SMITH, E.M.B. and SPREAFICO, F. (1987) *Immunotoxicology.* Dordrecht: Martinus Nijhoff.

BRIGGS, K.T., YOSHIDA, S.H. and GERSCHWIN, M.E. (1996) The influence of petrochemicals and stress on the immune system of seabirds. *Regul. Toxicol. Pharmacol.*, **23**, 145–155.

BURLESON, G.H. and DEAN, J.H. (1995) Immunotoxicology: past, present, and future. In: *Methods in Immunotoxicology* (Burleson, G.H., Dean, J.H. and Munson, A.E., eds), pp. 3–10. New York: Wiley-Liss.

BURREL, R., FLAHERTY, D.K. and SAUERS, L.J. (1992) *Toxicology of the Immune System. A Human Approach.* New York: Van Rostrand Reinhold.

DAVIES, G.E. (1983) Immunotoxicity: undesirable effects of inappropriate response. *Immunol. Today*, **4**, 1–2.

DAYAN, A.D., HERTEL, R.F., HESELTINE, E., KAZANTZIS, G., SMITH, E.M. and VAN DER VENNE, M.T. (1990) *Immunotoxicity of Metals and Immunotoxicology.* Proceedings of an International Workshop. New York: Plenum Press.

DEAN, J.H., PADARASINGH, N.L. and JERRELLS, T.R. (1979) Assessment of immunobiological effects induced by chemicals, drugs or food additives. I. Tier testing and screening approach. *Drug Chem. Toxicol.*, **2**, 5–17.

DESCOTES, J. (1986) *Immunotoxicology of Drugs and Chemicals*, 1st edition. Amsterdam: Elsevier Science.

DESCOTES, J. (1998) Regulating immunotoxicity evaluation: issues and needs. *Arch. Toxicol.*, **20**, S293–S299.

DESCOTES, J., NICOLAS, B., VIAL, T. and NICOLAS, J.F. (1996) Biomarkers of immunotoxicity in man. *Biomarkers*, **1**, 77–80.

FLÉCHET, M.L., MOORE, N., CHEDEVILLE, J.C., PAUX, G., BOISMARE, F. and LAURENT, P. (1985) Fatal epidermal necrolysis associated with isoxicam. *Lancet*, **ii**, 499.

GIBSON, G.G., HUBBARD, R. and PARKE, D.V. (1982) *Immunotoxicology.* Proceedings of the first International Symposium on Immunotoxicology. New York: Academic Press.

GÓMEZ DE LA CÁMARA, A., ABAITUA BORDA, I. and POSADA DE LA PAZ, M. (1997) Toxicologists versus toxicological disasters: Toxic Oil Syndrome, clinical aspects. *Arch. Toxicol.*, **suppl. 19**, 31–40.

GRAHAM, E.A. (1911) The influence of ether and ether anesthesia on bacteriolysis and phagocytosis. *J. Infect. Dis.*, **8**, 147–175.

Groupement Lyonnais de Lutte contre les Intoxications (1975) *Immunotoxicologie.* Proceedings of a meeting held in Lyon on 24 October 1974. Lyon: Editions Lacassagne.

KAMMÜLLER, M.E., BLOKSMA, N. and SEINEN, W. (1989) *Autoimmunity and Toxicology.* Amsterdam: Elsevier Science.

KIMBER, I. and MAURER, T. (1996) *Toxicology of Contact Hypersensitivity.* London: Taylor & Francis.

LUSTER, M.I., MUNSON, A.E., THOMAS, P.T., HOLSAPPLE, M.P., FENTERS, J.D., WHITE, K.L. *et al.* (1988) Development of a testing battery to assess chemical-induced immunotoxicity. National Toxicology Program's criteria for immunotoxicity evaluation in mice. *Fund. Appl. Toxicol.*, **10**, 2–19.

MEYLER, L. (1966) Cytostatics drugs. In: *Side Effects of Drugs*, Vol. V (Meyler, L., ed.), pp. 472–494. Amsterdam: Excerpta Medica.

NATIONAL RESEARCH COUNCIL (1992) *Biologic Markers in Immunotoxicology.* Washington DC: National Academy Press.

NEWCOMBE, D.S., ROSE, N.R. and BLOOM, J.C. (1992) *Clinical Immunotoxicology.* New York: Raven Press.

NILSSON, B.S. (1983) Adverse reactions in connection with zimeldine treatment: a review. *Acta Psychiatr. Scand.,* **68,** suppl. 308, 115–119.

RADFORD, S.G., LOCKYER, J.A. and SIMPSON, P.J. (1982) Immunological aspects of adverse reactions to althesin. *Br. J. Anaesth.,* **54,** 859–863.

SALAMA, A. and MUELLER-ECKHARDT, C. (1986) Two types of nomifensine-induced immune haemolytic anaemias: drug-dependent sensitization and/or autoimmunization. *Br. J. Haematol.,* **64,** 613–620.

SALVAGGIO, J.E. (1990) The impact of allergy and immunology on our expanding industrial environment. *J. Allergy Clin. Immunol.,* **85,** 689–699.

SANDLER, R.H. (1985) Anaphylactic reactions to zomepirac. *Ann. Emerg. Med.,* **14,** 171–174.

SELGRADE, M.J.K., COOPER, K.D., DEVLIN, R.B., VAN LOVEREN, H., BIAGINI, R.E. and LUSTER, M.I. (1995) Immunotoxicity – Bridging the gap between animal research and human health effects. *Fund. Appl. Toxicol.,* **24,** 13–21.

SNEL, J.J. (1903) Immunity and narcosis. *Klin. Wochen.,* **10,** 212–214.

Special Issue (1979) Immunotoxicology. *Drug Chem. Toxicol.,* **2,** 1 & 2.

Special Issue (1982) Target organ toxicity: immune system. *Environ. Health Perspect.,* **43,** February.

STRAIGHT, J.M., KIPEN, H.M., VOGT, R.F. and AMLER, R.W. (1994) *Immune Function Test Batteries for Use in Environmental Health Field Studies.* Atlanta: Agency for Toxic Substances and Diseases Registry.

STRICKER, B.H.C., DE GROOT, R.R.M. and WILSON, J.H.P. (1990) Anaphylaxis to glafenine. *Lancet,* **336,** 943–944.

SUNDWALL, A., ANDERSSON, B., BALLS, M., DEAN, J., DESCOTES, J., HAMMARSTRÖM, S., *et al.* (1994) Immunotoxicology and in vitro possibilities. *Toxicol. in vitro,* **8,** 1067–1074.

SWIFT, H.F. (1922) The action of sodium salicylate upon the formation of immune bodies. *J. Exp. Med.,* **36,** 735–760.

VAN LOVEREN, H., SRÁM, R. and NOLAN, C. (1997) *Environment and Immunity.* Air Pollution Epidemiology Reports Series. Report no. 11. Brussels: European Commission.

VIAL, T., NICOLAS, B. and DESCOTES, J. (1994) Drug-induced autoimmunity: experience of the French Pharmacovigilance system. *Toxicology,* **119,** 23–27.

VOS, J.G. (1977) Immune suppression as related to toxicology. *CRC Crit. Rev. Toxicol.,* **5,** 67–101.

VOS, J.G., YOUNES, M. and SMITH, E. (1995) *Allergic Hypersensitivities Induced by Chemicals.* Boca Raton: CRC Press.

2

Overview of the Immune System

It is outside the scope of this volume to provide a detailed description of the immune system. Readers are invited to refer to available immunology textbooks (suggested readings include: Abbas *et al.*, 1997; Janeway and Travers, 1996; Paul, 1993; Roitt *et al.*, 1996). Only general concepts of relevance for immunotoxicity evaluation from the perspective of non-immunologists will be briefly recalled in this chapter.

The primary function of the immune system in mammals and in lower species is the protection against microbial pathogens (bacteria, viruses, fungi and yeasts), and tumours. Every depletion of immune cells, immune dysregulation or functional defect is expected to facilitate the development of pathological events characterised by disturbances of the recognition mechanisms of the 'self' versus the 'non-self' and increased susceptibility of the host towards infections and neoplasia. An array of specialised (immunocompetent) cells, including white blood cells (leukocytes) and accessory cells are involved in immune responses. The renewal, activation, and differentiation of immunocompetent cells which are required to achieve a normal level of immune competence are under the control of many mechanisms with either redundant or conflicting outcome. Redundancy is indeed an essential feature of the immune system, so that the resulting consequences of functional immune alterations or defects caused by toxic exposures are often difficult to anticipate because of compensatory mechanisms. This situation is often referred to as the 'functional reserve capacity' of the immune system, even though no single or precise definition of the reserve capacity is widely accepted.

In order to play its role in the host's defences, the immune system utilises both specific and non-specific mechanisms, either simultaneously or independently. Specific immune responses include the humoral immunity (namely the production of antigen-specific antibodies by plasma cells derived from B lymphocytes) and the cell-mediated immunity involving T lymphocytes. In addition, closely intertwined intracellular relations ensure coordinated immune responses.

However, immune responses are not always beneficial: immune responses against self constituents of the host resulting in autoimmune diseases, and reactivity against an innocent antigen resulting in hypersensitivity reactions, are examples of deleterious immune responses.

Immunocompetent cells

Lymphocytes

Lymphocytes play an essential role in specific immune responses. They are divided into B and T lymphocytes based on the presence of surface markers, as they cannot be differentiated by cytological characteristics.

T lymphocytes constitute 55–75 per cent of circulating lymphocytes and express the CD2 and CD3 surface markers. They differentiate in the thymus where they express the antigen receptor TcR and divide into two subsets: T helper lymphocytes (60–70 per cent of T lymphocytes) which express CD4, and T cytotoxic lymphocytes which express CD8. T lymphocytes can also be differentiated into α/β TcR (by far the most numerous) or γ/δ TcR lymphocytes. $CD4^+$ T lymphocytes help B lymphocytes to divide, differentiate and produce antibodies. They produce cytokines which control the development of the leukocyte lineage, and they are required for the development of cytotoxic T lymphocytes, the activation of macrophages, etc. Based on the profile of the cytokines they release, Th1 (interleukin-2 (IL-2), interferon gamma (IFN-γ), and tumour necrosis factor beta (TNF-β)) and Th2 (IL-4, IL-5 and IL-10) lymphocytes have been identified. T cytotoxic lymphocytes kill cells infected by viruses, or tumour cells. The majority of T cytotoxic lymphocytes are $CD8^+$ T cells.

B lymphocytes (10–20 per cent of peripheral blood lymphocytes) first develop in the foetal liver, then the bone marrow. They can be identified by the presence of surface immunoglobulins. At the final stage of their maturation, they become plasma cells, which synthesise and release antigen-specific antibodies.

During maturation, each lymphocyte synthesises a receptor for the antigen with a unique specificity. An antigen can specifically bind to a small number of lymphocytes. The whole range of antigen-binding specificities is called the antigenic repertoire of lymphocytes. Memory lymphocytes are either B or T lymphocytes with a long half-life, which trigger an amplified and accelerated response to a given antigen after a first contact.

Nul (nonB-, nonT-) cells account for less than 15 per cent of peripheral blood mononuclear cells. They have no receptor for the antigen and might represent a particular cell lineage.

Antigen-presenting cells

Antigen-presenting cells (APCs) have the capacity to process and present antigens to lymphocytes in a way they can be recognised. B lymphocytes recognise native antigens, whereas T lymphocytes recognise antigenic peptides associated with major histocompatibility complex (MHC) molecules. For antigen presentation to T cells, APCs must internalise, then process the antigen into small peptidic fragments to be subsequently expressed onto the membrane surface in association with MHC class I molecules when $CD8^+$ T cytotoxic lymphocytes are involved and MHC class II molecules when $CD4^+$ T helper lymphocytes are involved.

The main types of APCs are Langerhans cells (recirculating cells from the skin which express CD1 and MHC class II molecules), dendritic cells (residing cells which are distributed throughout most of the tissues), and macrophages. A variety of cells, such as thyroid cells, also have the capacity to express MHC class II molecules under the influence of IFN-γ released by activated $CD4^+$ T cells, to become APCs.

Table 2.1 Main CD antigens and cells expressing these antigens

CD antigen	Expressing cells
CD1	Cortical thymocytes, Langerhans cells, dendritic cells
CD2	T cells, thymocytes, NK cells
CD3	T cells, thymocytes
CD4	Helper T lymphocytes
CD5	Thymocytes, some T and B cells
CD8	Cytotoxic T lymphocytes
CD11a	Lymphocytes, granulocytes, monocytes, macrophages
CD11b	Natural cytotoxic cells
CD16	Neutrophils, NK cells, macrophages
CD18	Leukocytes
CD19	B cells
CD21	Mature B cells, dendritic cells
CD23	Mature B cells, activated macrophages, eosinophils, follicular dendritic cells
CD25	Activated T cells, B cells, monocytes
CD28	Some T cells, activated B cells
CD31	Monocytes, platelets, granulocytes, endothelial cells
CD40	B cells, monocytes, dendritic cells
CD44	Leukocytes, erythrocytes
CD45	Memory T cells
CD56	NK cells
CD72	B cells

Phagocytes

All phagocytes derive from multipotent medullary stem cells. Neutrophil leukocytes, with two or three nuclear lobules, account for over 70 per cent of leukocytes, and over 95 per cent of polynuclear blood cells derive from the granulocytic cell lineage. They adhere to blood vessel walls and leave blood under the influence of chimiotactic factors and further migrate to tissues where they exert their phagocytic and bactericidal functions. Neutrophil leukocytes store two types of granules: the so-called primary granules, which release lytic enzymes, such as lysozyme, myeloperoxidase-derived substances, such as hydrogen peroxide, and cationic proteins; and the secondary granules, which contain enzymes, such as lactoferrin and collagenase.

Mononuclear phagocytes include circulating monocytes (5 per cent of leukocytes) and tissue macrophages derived from monocytes. Macrophages have a wide morphological and functional polymorphism according to the tissue location (Kupffer cells in the liver, alveolar and peritoneal macrophages, etc.). In contrast to polynuclear phagocytes, mononuclear phagocytes are potent antigen-presenting cells in addition to their role in phagotytosis. Macrophages, when activated, release cytokines, such as IL-1, IL-6 and TNF-α.

Accessory cells

Additional cells are involved in varied immune responses. NK ('natural killer') cells have morphological features resembling those of large granular lymphocytes. They directly kill target cells, such as cells infected by viruses or tumour cells without prior sensitisation.

In addition to neutrophils, polynuclear leukocytes include eosinophils and basophils. Eosinophils, which account for 2–5 per cent of leukocytes, kill parasites. Less than 0.5 per cent of leukocytes are basophils which store mediators, such as histamine, to be abruptly released during anaphylactic and pseudo-allergic hypersensitivity reactions.

Mast cells are found in many sites lining blood vessels. Similar to basophils, they store pro-inflammatory mediators, such as histamine and the platelet-activating factor (PAF), which are released after binding of antigen-specific IgE antibodies to high affinity membrane receptors (FcεRI). Pro-inflammatory mediators, such as prostaglandins and leukotrienes, are also released with a delayed time course because they are not stored in granules as histamine, but must first be synthesised to be released. Mast cells are divided into connective tissue mast cells, the most numerous subpopulation, which contain large amounts of histamine and heparin, and mucosal mast cells found in the lung and the gut.

The lymphoid system

Immunocompetent cells can be found throughout the body, but they preferentially concentrate into specific anatomical structures, the lymphoid organs. Lymphoid organs are divided into central (or primary) lymphoid organs, and peripheral (secondary) lymphoid organs, with different roles on the maturation of immunocompetent cells and the implementation of immune responses.

Central lymphoid organs

Central lymphoid organs are lympho-epithelial structures which develop early during organogenesis, independently of any antigenic stimulation. They ensure the production and/or the maturation of lymphocytes.

The thymus

The thymus is the main central lymphoid organ. Lobules, the thymic functional units, are surrounded by epithelial cells and comprise of the cortical area with many immature thymocytes, the vast majority of which die inside the thymus (following apoptosis), and the medullary area with mature (CD4$^+$ CD8$^-$ or CD4$^-$ CD8$^+$) lymphocytes ready to leave the thymus. Dendritic cells as well as macrophages are also found in the medulla. The epithelial micro-environment and thymic hormones ensure the maturation of T lymphocytes. The thymus normally shrinks during life.

The bone marrow

The bone marrow is the human equivalent of the bursa of Fabricius present in birds. In the bone marrow, multipotent progenitor cells are produced, which later give birth to all immunocompetent cells. B lymphocytes mature in the bone marrow.

Peripheral lymphoid organs

Antigens and immunocompetent cells come into contact in peripheral lymphoid organs because of the rich lymphatic vascularisation. These organs develop late during gestation and reach optimal development only after repeated antigenic stimulation. They contain

small clusters of B lymphocytes: the lymphoid follicles, which are called primary follicles prior to antigenic stimulation; and secondary follicles with a germinative centre
including lymphoblasts, following antigenic stimulation.

The spleen

The spleen acts as a filter in the blood circulation. It consists of the red pulp (lined with
macrophages) and the white pulp made of peri-arterial sheaths (the thymo-dependent
zone) and follicles of B lymphocytes (the thymo-independent zone).

Lymph nodes

Lymph nodes comprise a cortical area in which lymphoid follicles with B lymphocytes
can be found, a paracortical area with mainly T lymphocytes, and a medullary area with
few lymphocytes, but many macrophages and plasma cells. Afferent and efferent lymphatic vessels are found within lymph nodes.

Specialised lymphoid tissues

Mucosa-associated lymphoid tissue (MALT) can be found in the gut (gut-associated
lymphoid tissue or GALT), corresponding to Peyer's patches, the bronchopulmonary
tract (bronchi-associated lymphoid tissue or BALT), and the upper airways. The exact
physiological role of specialised lymphoid tissues is still ill-understood.

Immune responses

Immune responses result from the combined effects of immunocompetent cells and
soluble factors. They are either antigen-specific ('specific' or 'adaptive' immunity) or
non-antigen-specific ('natural' immunity). Antigen-specific immune responses are further
divided into humoral and cell-mediated responses.

Humoral immunity

After contact with an antigen, the immune system can produce antigen-specific antibodies.

Antibodies

Antibodies are heterodimeric glycoproteins (immunoglobulins or Ig) with five classes
found in humans and most mammals, namely IgG, IgM, IgA, IgE and IgD. They consist
of two heavy (H) and two light (L) chains linked by disulphide bridges of varying numbers and locations according to the Ig class or subclass. Disulphide bridges are critical to
determine the tertiary structure of the immunoglobulins.

Both chains contain variable (V) and constant (C) regions, which differ due to the
variability of the amino acid sequence. In addition, disulphide bridges define domains
(VH, VL, C1L, C1H, C2H, C3H) in the aminoacid sequence. The V regions are divided
into framework (Fr) and complementarity-determining regions (CDRs). The latter display
the greatest aminoacid variability and act as antigenic determinants (idiotopes). Multiple
independent idiotopes in a single immunoglobulin molecule are called idiotypes. Heavy

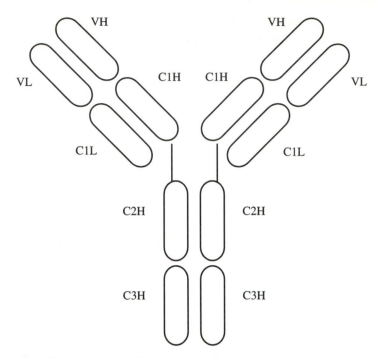

Figure 2.1　Simplified structure of immunoglobulins

chains differ according to the Ig class: gamma, mu, alpha, epsilon or delta chains, but there are only two types of light chains, namely kappa and lambda, whatever the Ig class. The N-terminal end of the chain is quite different between chains; this is the variable end which supports the antibody specificity. By contrast, the C-terminal end is much more consistent.

The antibody site is formed by the variable domains of the light and heavy chains. Each Ig molecule can be divided into two Fab fragments, which bind the antibody and one Fc fragment, which is involved in various biological functions, such as complement activation, opsonisation and placental transfer.

Primary and secondary antibody responses

In lymphoid organs, B lymphocytes when stimulated by the antigen, are transformed into plasma cells, which synthesise and release antibodies. The number of possible antibodies, or repertoire, is extremely large. There are genes coding for constant ends and genes coding for variable ends of immunoglobulins, which are located on different chromosomes. An extremely large number of immunoglobulins can be formed following gene rearrangements.

After a first contact with the antigen (primary response), antibody production develops in three phases. The latency period usually lasts 3–4 days before the first antibodies are produced, but the duration depends on the route of antigen administration, the dose and the nature of the antigen. Antibody titres subsequently grow exponentially and finally decrease either totally or down to a residual level. During a primary response, the first antibodies to be detected are IgM, then IgG. After a second contact (secondary response), antibody production is earlier and more marked. Antibodies are IgG of a higher affinity for the antigen.

Antibody production requires that B lymphocytes expressing antigen-specific membrane immunoglobulins are activated resulting in increased numbers of antigen-responding cells (clonal expansion). While they proliferate, B lymphocytes are transformed into lymphoblasts, which give rise to long-living memory B lymphocytes, and to plasma cells, which produce antibodies.

Cell-mediated immunity

The immune response can also involve strictly cellular mechanisms without the production of antigen-specific antibodies.

Cell–cell interactions

Cell-mediated immune responses are the result of complex cell–cell interactions including the processing, then the presentation of the antigen to T lymphocytes by macrophages under the strict genetic control of major histocompatibility complex class I and II molecules. A number of soluble or cellular molecules interplay in cell–cell interactions: examples of such molecules are the cell adhesion molecules, such as the selectins (ICAM-1/LFA-1, VCAM-1) and the interleukins.

Interleukin-1 (IL-1) is released by stimulated macrophages to activate CD4$^+$ helper T lymphocytes, whereas IL-2 is released by helper T lymphocytes and augments the expression of its own IL-2 receptors on other helper T lymphocytes, which in turn release various cytokines, which activate other immunocompetent cells, such as CD8$^+$ cytotoxic T lymphocytes.

Cytokines

Cytokines which are released by leukocytes and other cells, play a pivotal role in immune responses. They exert many and often redundant activities (the so-called pleiotropic effects of cytokines). Cytokines include colony-stimulating factors (CSF), such as granulocyte-colony stimulating factor (G-CSF), granulocyte/macrophage-colony stimulating factor (GM-CSF), IL-3, interferons alpha, beta and gamma, the tumour necrosis factors (TNF-α and β), interleukins (from IL-1 to IL-18 . . . at least for the time being!).

Non-specific immune responses

Phagocytosis is the major mechanism of non-specific defences to eradicate foreign bodies and microbial pathogens. Phagocytes move close to their targets under the influence of chemotactic substances (chemotaxis), which adhere to their surface, either non-specifically (lectins) or following opsonisation (the main opsonins are IgG and C3b). Phagocytes engulf their prey by invagination of their cell membrane (leading to phagosomes) and release lyzosomial enzymes (lysozyme, cationic proteins, proteases, peroxidase) and free radicals.

Cytotoxicity can also be a non-specific effector mechanism, such as cytotoxicity induced by NK cells.

External influences on the immune system

Importantly, the immune system is not physiologically isolated.

Table 2.2   Sources, targets and main effects of major cytokines

Cytokine	Source	Target	Effects
IL-1	Macrophages, fibroblasts	Lymphocytes, macrophages	Fever, activation of T lymphocytes and macrophages
IL-2	T lymphocytes	T lymphocytes	Proliferation of T lymphocytes
IL-3	T lymphocytes, thymic cells	Stem cells	Stimulation of haemopoiesis
IL-4	T lymphocytes, mast cells	B lymphocytes	Increased IgE production
IL-5	T lymphocytes, macrophages	B and T lymphocytes	Growth and differentiation of eosinophils
IL-6	Macrophages, T lymphocytes	T lymphocytes, B lymphocytes, hepatocytes	Growth and differentiation of B and T lymphocytes Synthesis of acute phase proteins
IL-7	Bone marrow	Lymphocytes	Growth of pre-B and pre-T lymphocytes
G-CSF	Fibroblasts	Neutrophils	Growth and stimulation of neutrophils
GM-CSF	Macrophages, T lymphocytes	Granulocytes, macrophages	Growth and stimulation of granulocytes and macrophages
TNF-α	Macrophages	Natural killer cells	Inflammation, activation of endothelial cells
TNF-β	T lymphocytes, B lymphocytes		Cell death, activation of endothelial cells
IFN-α	Leukocytes	Antiviral effects	Increased expression of MHC class I molecules
IFN-β	Fibroblasts	Antiviral effects	Increased expression of MHC class I molecules
IFN-γ	T lymphocytes	Natural killer cells	Activation of macrophages, increased expression of MHC class II molecules

Neuroendocrine system

The immune system is under the continuous influence of the central and autonomic nervous system (Ader *et al.*, 1990) as well as the endocrine glands. Stress, either physical or psychological, is a critical factor to be considered (Plotnikoff *et al.*, 1991), particularly in the context of immunotoxicity evaluation. For instance, strenuous sports (Nash, 1993), such as marathon running, or major depression (Maes, 1995) have been shown to induce immunodepression. The role of psychological factors in the resistance of laboratory animals to experimental infections, or in the response of human beings to disease has been emphasised and this led to the identification of psychoneuroimmunology as a new sub-discipline (Ader and Cohen, 1993). There is experimental evidence that behavioural conditioning can interfere with the immune response. Neuromediators, such as noradrenaline, acetylcholine and serotonin, have immunomodulatory activities (Dunn, 1995; Madden and Felten, 1995). The immune effects of hormones are not restricted to the immuno-suppressive cortisol (Frier, 1990): oestrogens, progesterone and androgens, prolactin, growth hormone, and thyroid hormones modulate the immune response.

SHOPPERS FOOD & PHARMACY
4720 Cherry Hill Road
College Park, MD 20740
STORE MANAGER: Mike Morris
301 345 5096
www.shoppersfood.com

TAX $0.02
**** TOTAL ****   15.00
Cash

CHANGE

Thank you for shopping with us today!
Please visit our FAMILY PHARMACY
Pharmacy phone number: 301 345

Nutrition and aging

Nutrition and aging should not be ignored. Malnutrition is probably the first cause of immunodepression in human beings worldwide, and deficiency in various vitamins, such as vitamins A, C and E, or trace elements, such as selenium and zinc, has been shown to be associated with impaired immune responses (Chandra, 1993). Immune responsiveness tends to decrease with age: the thymus normally shrinks during life, but the real impact of immune senescence is still debated (Voets *et al.*, 1997). Because non-clinical immunotoxicity evaluation studies are conducted in young adult healthy animals, the adverse influence of nutrition and aging in specific groups of the population should be considered in immunotoxicity risk assessment.

Influence of the immune system on other systems

In turn, the immune system can influence other physiological functions. IL-1 and inter-feron-α are somnogenic and can alter the EEG. They also both decrease locomotor activity in laboratory animals. Several cytokines have neurotransmitter activity and their possible role in psychiatric diseases is an area of growing research. IL-1 (formerly called the endogenous pyrogen) has a marked influence on central thermoregulation, whereas IL-6 is involved in the acute-phase response and exerts inhibitory effects on cytochrome P450-dependent drug metabolism (Lian Chen *et al.*, 1994). The impact of the immune system on other physiological functions is still poorly recognised, although it may account for adverse effects caused by immunomodulating compounds.

Conclusion

This brief overview of the immune system shows how complex and intricated immune responses can be, due to the interplay of many redundant or conflicting mechanisms. In addition, other biological systems, such as the second messengers or the complement pathway, are involved. Therefore, difficulties are often encountered when interpreting the results of immunotoxicity evaluation studies to determine their relevance and mechanism.

References

ABBAS, A.K, LICHTMAN, A.H. and POBER, J.S. (1997) *Cellular and Molecular Immunology.* 3rd edition. Philadelphia: W.B. Saunders.

ADER, R. and COHEN, N. (1993) Psychoneuroimmunology: conditioning and stress. *Annu. Rev. Psychol.*, **44**, 53–85.

ADER, R., FELTEN, D. and COHEN, N. (1990) Interactions between the brain and the immune system. *Annu. Rev. Pharmacol. Toxicol.*, **30**, 561–602.

CHANDRA, R.K. (1993) Nutrition and the immune system. *Proc. Nutr. Soc.*, **52**, 77–84.

DUNN, A.J. (1995) Interactions between the nervous system and the immune system. Implications for psychopharmacology. In: *Psychopharmacology: the Fourth Generation of Progress* (Bloom, F.E. and Kupfer, D.J., eds) pp. 719–731. New York: Raven Press.

FRIER, S. (1990) *The Neuroendocrine–Immune Network.* Boca Raton: CRC Press.

JANEWAY, C.A. and TRAVERS, P. (1996) *Immunobiology.* 2nd edition. London: Current Biology Ltd, Garland Publishing.

LIAN CHEN, Y., LE VRAUX, V., LENEVEU, A., DREYFUS, F., STHENEUR, A., FLORENTIN, A., *et al.* (1994) Acute-phase response, interleukin-6, and alteration of cyclosporine pharmacokinetics. *Clin. Pharmacol. Ther.*, **55**, 649–660.

MADDEN, K.S. and FELTEN, D.L. (1995) Experimental basis for neural-immune interactions. *Physiol. Rev.*, **75**, 77–106.

MAES, M. (1995) Evidence for an immune response in major depression: a review and hypothesis. *Prog. Neuro-Psychopharmacol. Biol. Psychiatr.*, **19**, 11–38.

NASH, M.S. (1993) Exercise and immunology. *Med. Sci. Sports Exerc.*, **26**, 125–127.

PAUL, W.E. (1993) *Fundamental Immunology*, 3rd edition. New York: Raven Press.

PLOTNIKOFF, N.P., FAITH, R.E., MURGO, A.J. and WYBRAN, J. (1991) *Stress and Immunity.* Boca Raton: CRC Press.

ROITT, I., BROSTOFF, J. and MALE, D. (1996) *Immunology*, 4th edition. London: Gower Medical.

VOETS, A.J., TULNER, L.R. and LIGTHART, G.J. (1997) Immunosenescence revisited: does it have any clinical significance? *Drugs & Aging*, **11**, 1–6.

3

Immunosuppression and Immunodepression

Immunosuppression is the area of immunotoxicology, the adverse consequences of which are best recognised on human health; this is largely due to the introduction of potent immunosuppressive drugs into the clinical setting in the 1960s, resulting in the early description of adverse reactions induced by these new treatments, but also to the focused and continuing interest of many immunotoxicologists on immunosuppression.

The two major adverse consequences of immunosuppression are impaired host resistance against microbial pathogens leading to infectious complications, and more frequent neoplasias, particularly lymphomas. A question, to which so far has been paid limited attention, is whether, and if so, to what extent, mild to moderate impairment of the immune responsiveness (namely immunodepression) can result in similar adverse health consequences.

Impaired resistance against microbial pathogens

Most primary (or congenital) and secondary (acquired) immune deficiencies have been shown to be associated with infectious complications, whatever the type of immune defect involved, namely specific or non-specific. Illustrative examples from the general medical practice include infants born with a severe immune defect, and cancer or AIDS patients, all of whom develop frequent, severe and potentially lethal infections (Rubin and Young, 1994).

Similar examples can be found in the area of toxicology. Infectious complications have long been described in patients treated with long-term and/or high-dose corticosteroid therapy (Kass and Finland, 1953; Thomas, 1952), immunosuppressants or cytotoxic anticancer drugs (Davies, 1981). It is therefore not surprising that infections are so common in transplant patients (Nicholson and Johnson, 1994; Singh and Yu, 1996). Even though the incidence of infectious diseases has seldom been studied in human beings exposed to occupational or environmental immunotoxicants, the few available data show a similar trend towards more frequent, even though clinically unremarkable, infections in individuals exposed to immunodepressive chemicals, such as biphenyls (Kuratsune *et al.*, 1996), addictive drugs (Pillai *et al.*, 1991) or heavy metals (Zelikoff and Thomas, 1998).

Involved pathogens

Although infections in immunocompromised patients can be caused by any pathogen, several microbial species are more often found.

Bacterial infections are preferentially due to staphylococci (*Staphylococcus aureus* and *S. epidermidis*), streptococci (*Streptococcus pneumoniae*), *Escherichia coli*, *Pseudomonas aeruginosa*, *Listeria monocytogenes*, *Haemophilus influenzae*, Nocardia, and Mycobacteria (Donnelly, 1995). *L. monocytogenes* is an intracellular pathogen that occurs in patients with a defect in cellular immunity. Meningitis is the most common complication, but bacteraemia, brain abscess, meningo-encephalitis, or pneumonia may also develop. Nocardiosis is primarily an infection of the lung, but dissemination to the skin, the central nervous system, and the skeletal system is also common. Apart from tuberculosis, mycobacterial infections involve atypical mycobacteria, such as *M. kansasii*, *M. chelonei*, and *M. haemophilum*.

Herpes viruses and cytomegalovirus are the most common viral agents (Griffiths, 1995). Cytomegalovirus infections are frequent, particularly in renal transplant patients, and clinically the disease can range from asymptomatic viral shedding to life-threatening disseminated disease. Herpes simplex virus infections are also frequent, and often due to reactivation of a dormant disease. Other viruses that cause infections in immuno-compromised patients include the Epstein–Barr virus resulting in asymptomatic elevation in serum antibody titres or uncommonly in disseminated lymphoproliferative disorder, hepatitis B, C and delta viruses, respiratory syncytial virus, varicella-zoster and influenza viruses.

Fungal infections are a major cause of morbidity and mortality in immunocompromised patients. Infections due to *Candida* and *Aspergillus* are the most frequent. The most common site of *Candida* infections is the oral mucosa, but disseminated infections may occur. *Aspergillus* infections predominate in the lung. Most parasitic infections in immuno-compromised patients are due to *Pneumocystis carinii*, *Toxoplasma gondii*, *Strongyloides stercoralis* and *Cryptosporidiae*.

Clinical features

Microbial infections can be located at any site in immunocompromised patients, even though several sites are more frequently involved (Rubin and Young, 1994).

Broncho-pulmonary infections are the most frequent infectious complications in immunocompromised patients (Frattini and Trulock, 1993). As the respiratory tract is the major route of entry for microbial pathogens, it is not surprising that infections of the upper airways and the lung are so common. In addition, many specific and non-specific immunological mechanisms are involved in the protection of airways against microbial pathogens. Cytomegalovirus infections are particularly frequent (Rubin, 1993). They are often silent and found in up to 90 per cent of renal transplant patients. Other commonly found lung pathogens include *Bacillus tuberculosis* and pyogenic bacteria, whereas Gram negative bacteria (e.g. *Pseudomonas aeruginosa*) and fungi (*Candida*, *Nocardia*, *Aspergillus*) are less common.

The digestive tract is the first line of defence against oral pathogens. Any dysfunction of specific and non-specific defence mechanisms can result in more frequent and/or more severe gastrointestinal infections with often atypical clinical features and prolonged or recurring outcome (Bodey and Fainstein, 1986). Even though broncho-pulmonary infections

Table 3.1 Typical infectious complications associated with the main types of immune function defect

Immune defect	Infectious complications	Causative pathogens
Cell-mediated immunity	Systemic or localised opportunistic infections of the lung, nervous central system, skin; malignant viral infections; chronic diarrhoea	*Listeria monocytogenes*, *Bacillus tuberculosis*, atypical mycobacteria, *Aspergillus*, *Candida*, *Pneumocystis carinii*, herpes virus, measles *Cryptosporidiae*, *Giardia*
Humoral immunity	Broncho-respiratory infections	*Haemophilus*, streptococci, staphylococci, *Klebsiella*
	Viral infections	Hepatitis B virus, echovirus, rotavirus
	Gastrointestinal infections	*Giardia, Lamblia*
Phagocytosis	Severe and often relapsing infections of the skin, lymph nodes and lung	staphylococci, *Salmonella*, *Klebsiella, Pseudomonas*
Complement	Severe and often relapsing infections of the skin	streptococci, staphylococci, *Neisseria*

are more frequent, gastrointestinal infections (e.g. chronic diarrhoeal syndrome) may predominate in some patients. Associated risk factors, such as atrophy of the gastric mucosa, increased gastric acid secretion, and alterations of the intestinal flora, contribute to the development of gastrointestinal infections in immunocompromised patients. Unexpectedly, selective IgA deficiencies are seldom associated with gastrointestinal infections, even though IgA are the predominant class of immunoglobulins in digestive secretions. The pathogens more often involved in gastrointestinal infections include *Salmonella*, *Aeromonas*, rotavirus, adenovirus and cytomegalovirus.

Beside broncho-pulmonary and gastrointestinal infections, which are by far the most frequent infections in immunocompromised patients, any other site can be infected. However, these infectious complications usually have atypical clinical features, most notably infections of the central nervous system (brain abscess, encephalitis), or severe and recurring infections of the skin. By contrast, urinary infections are relatively uncommon. Finally, some infectious complications may remain silent, or present as isolated fever.

Immune changes and infectious complications

Every immunocompromised patient is at a greater risk of developing more frequent and/or more severe infectious complications, and the type of infection is at least partially linked to the type of immune changes evidenced in these patients.

Data from patients with congenital immune deficiencies are particularly illustrative in this respect (Rosen *et al.*, 1995). Patients with congenital defects of the cell-mediated immunity develop severe infections due to pathogens with limited virulence, such as viruses (herpes virus, measles virus), intracellular pathogens (*Listeria monocytogenes*, *Bacillus tuberculosis*, atypical mycobacteria), fungi and parasites (*Aspergillus, Candida*,

Pneumocystis carinii). Progressively lethal infections may develop following vaccination using a living attenuated pathogen, such as poliomyelitis vaccine or BCG (MacKay *et al.*, 1980). Chronic diarrhoeal syndromes due to cryptosporidiae, *Giardia* and rotavirus are extremely frequent.

In patients with congenital defects of humoral immunity, infections are mainly due to *Haemophilus influenzae*, *Streptococcus pneumoniae*, *Staphylococcus aureus* or *Klebsiella pneumoniae*. Infections of the broncho-respiratory tract are the most common. Viral infections due to hepatitis B virus, echovirus, or rotavirus are also described. Gastrointestinal infections to *Giardia* or *Lamblia* are common.

Congenital defects of phagocytosis or chemotaxis are characterised by frequent and severe bacterial infections to staphylococci, *Salmonella*, *Klebsiella*, *Serratia and Pseudomonas*, with the skin, the lymph nodes and the lung as the most common sites.

Congenital defects of the complement system are also associated with systemic or localised (skin and upper airways) bacterial infections to pyogenic pathogens, with a prolonged and recurring evolution. Extremely severe infections to *Neisseria* have been reported.

The wide range of mechanisms involved to defend the body against microbial pathogens account for the varied clinical features of infectious complications (Revillard, 1990). Besides non-specific mechanisms which can be very effective in many instances, specific mechanisms are also involved. Specific mechanisms are manifold. Among antibacterial mechanisms are antibody neutralising bacteria, cytotoxic antibodies and cytokine production. Antiviral mechanisms include antibodies blocking virus adsorption on or lysis of infected cells, and ADCC (antibody-dependent cell-mediated cytotoxicity), cytotoxic T lymphocytes, interferons and NK cells, whereas cytolytic antibodies, lymphokines and eosinophils are the main antiparasitic defence mechanisms.

Antimicrobial resistance in immunotoxicological evaluation

Available clinical data show how important antimicrobial resistance is for maintained good health. Because an enormous variety of mechanisms, either redundant or conflicting, can be simultaneously or independently involved, it is often difficult to predict the resulting impact of given immune changes on antimicrobial resistance. The development of infectious complications in exposed individuals is indeed much more convincing evidence of the immunotoxic role of the chemical exposure under consideration, than the finding of immune function changes and still more so, than merely descriptive alterations of the normal architecture of selected lymphoid organs. The use of experimental infection models is therefore logical in immunotoxicity evaluation, but these models must be carefully selected to evidence the adverse consequences of unexpected immunosuppression related to chemical exposure. This question is considered later in this volume (see Chapter 13).

Another approach to assess the risk of infectious complications associated with immunosuppression is the use of epidemiological studies. Epidemiological studies are particularly helpful to detect an increased incidence of otherwise clinically unremarkable complications in exposed groups of the population. In addition, they can help identify causative or facilitating factors (at-risk individuals). However, a major difficulty today is the lack of reliable biomarkers of immunotoxicity so that it is often impossible to provide direct evidence of a causal link between the immunosuppressive or immunodepressive influence of a given chemical exposure and the observed increased incidence of infections.

Increased incidence of cancer

Lymphomas and acute leukaemias have been repeatedly reported to be more frequent in severely immunocompromised patients. Data in patients with congenital or acquired immune defects, such as the Wiskott–Aldrich syndrome, ataxia–telangiectasia, and AIDS, support this conclusion. Other lines of evidence come from patients with second malignancies, and from organ transplant patients.

Second malignancies and chemotherapy

An early indication that a relation might exist between immunosuppression and neoplasia was the identification of 'second malignancies' in cancer patients treated with prolonged chemotherapy (Penn, 1994). Second malignancies are cancers which develop in patients with a history of a previously diagnosed, but unrelated, first cancer.

Approximately 1 per cent of patients develop second malignancies within ten years after completion of chemotherapy, and 3 per cent within 20 years. Acute leukaemias (particularly acute myeloid leukaemias), non-Hodgkin's lymphomas, and solid tumours (carcinoma of the skin, lung, breast, colon and pancreas) are the most common second malignancies. A carcinogenic effect of chemotherapy can partly account for second malignancies, as alkylating agents, which are the most commonly involved anticancer agents in this context, are well-recognised genotoxicants. Another uncertainty derives from the comparison of patients with a history of previous malignancy to the general population, as a bias can be found in the possible, even though theoretical, role of individual predisposition to malignancy.

In any event, as some second malignancies, particularly lymphomas, have been shown to develop within a few years after completion of chemotherapy, a typical genotoxic mechanism is unlikely to be involved. As most chemotherapeutic agents are also potent immunosuppressants (Ehrke *et al.*, 1989), the role of immunosuppression in second malignancies was suggested.

Cancer in renal transplant patients

Interestingly, data in renal transplant patients support the hypothesis that immunosuppression is associated with more frequent neoplasias (Kinlen *et al.*, 1980). Clinical studies and cancer registries in transplant patients found that cancer develops in 1 to 15 per cent of organ transplant patients with an average of five years post-transplantation. Cancers of the skin and lips are the most frequently reported malignancies. They account for 30 per cent of all malignancies in transplant patients, and are observed in up to 18 per cent of these patients. The incidence of squamous cell carcinoma is about 250 times that of the general population. Lymphomas account for 14 to 18 per cent of all malignancies in transplant patients, and the incidence of lymphomas is approximately 40 times greater than in the general population. These lymphomas are non-Hodgkin's lymphomas in 95 per cent of cases. Kaposi's sarcoma which accounts for 4 to 6 per cent of malignancies in transplant patients, is 400 to 500 times more frequent than in the general population.

Cancer is more frequent in organ transplant patients, whatever the immunosuppressive agent(s) used, thus ruling out the possible role of genotoxicity, as most novel immunosuppressants, such as cyclosporine or tacrolimus (FK506), are not genotoxic. The incidence of malignancies was estimated to be 100 times greater in azathioprine-treated patients than in control populations. The most common cancers in these patients were non-Hodgkin's lymphomas, squamous cell cancers of the skin and primary liver tumours. Cyclosporine was also associated with a greater incidence of neoplasias, particularly lymphomas and Kaposi's sarcomas. Lymphoproliferative disorders have also been described in transplant patients treated with the monoclonal antibody OKT3, the macrolide derivative rapamycin, methotrexate and even corticosteroids, but quite uncommonly (Li *et al.*, 1990).

Cancer in other groups of patients

Lymphomas and acute leukaemias have also been noted in non-cancerous and non-transplant patients treated with prolonged immunosuppressive regimens, such as rheumatoid arthritis patients. However, these findings have to be carefully interpreted as malignant haemopathies were suggested to be more frequent in rheumatoid patients, even though they are not treated with immunosuppressive drugs (Kamel *et al.*, 1995).

Few data are supportive of an increased incidence of cancers in humans exposed to immunosuppressive chemicals despite growing concern fuelled by the AIDS epidemic (Margolick and Vogt, 1991). In laboratory animals, a number of carcinogenic chemicals, such as benzene, benzo[a]pyrene, benzidine, vinyl chloride, dimethylbenz[a]anthracene, dioxin, ethylcarbamate, and methylcarbamate, are also immunosuppressive, but it remains to establish whether these chemicals are also immunosuppressive in exposed human beings.

A few examples can illustrate current uncertainties. Between 1973 and 1974, over 500 farms in Michigan were contaminated by brominated biphenyls, accidentally mixed to cattle feed. Thirty-four per cent of exposed farmers were found with decreased T lymphocyte numbers and impaired proliferative responses to mitogens (Silva *et al.*, 1979). Fifteen years later, the incidence of cancers was suggested to be greater in exposed farmers compared to matched controls from Wisconsin, but these preliminary findings were never confirmed. Benzene is clearly toxic to the blood-forming cells of the bone marrow, causing aplastic anaemias, acute myelogenous leukaemias and possibly lymphomas. The toxic effects of benzene are due to metabolites rather than the parent molecule, and key metabolites in that respect, such as parabenzoquinone and hydroquinone, were shown to be markedly immunosuppressive (Smialowicz, 1996). Another interesting issue is the possible association between pesticide exposures and lymphomas. A number of epidemiological studies found that rural areas in which inhabitants are logically assumed to be more exposed to pesticides than in cities, are at a greater risk of developing lymphomas. However, these findings have not been confirmed by a number of other epidemiological studies, and the immunosuppressive activity of most pesticides has not been established so far, either in laboratory animals or in human beings (Vial *et al.*, 1996). As with infectious complications of immunosuppression, a major difficulty is the lack of reliable biomarkers of immunotoxicity to establish causal link between the immunosuppressive influence of chemical exposures and the development of cancer.

Mechanisms

Whatever the actual magnitude of the incurred risk, marked and prolonged immuno-suppression is associated with an increased incidence of malignancies.

Tumour antigens have long been identified on murine neoplastic cells, but their presence was evidenced on human neoplastic cells only very recently, hence the debate on whether the concept of the immunosurveillance of cancer as proposed by Paul Ehrlich in the early years of this century was valid. This concept served as a rationale for extensive research efforts to design cancer immunotherapy in the 1960s–1970s, but these efforts proved largely unsuccessful. The recent discovery of human tumour antigens fuelled the renewed interest in the design of anticancer vaccines. However, it is no longer accepted that specific immune responses can be involved in the induction of cancer by physical or chemical agents.

NK cells are considered to play a major role in the protection against cancer. Interestingly, the majority of polycyclic aromatic hydrocarbons are carcinogenic and the majority of those which are carcinogenic, induce a decrease in NK cell activity.

Another important issue is the interaction between immunosuppression and oncogenic viruses. In fact, a number of immunosuppression-related malignancies are induced by viruses, such as the Epstein–Barr virus and human papilloma viruses. Activation of the Epstein–Barr virus was evidenced in patients with lymphoproliferative disorders associated with azathioprine or cyclosporine treatments (Birkeland *et al.*, 1995). One attractive, and widely accepted, hypothesis is that dormant Epstein-Barr virus infection is facilitated by impaired T cell function in relation to immunosuppressive therapy. Interestingly, total recovery was noted in a few patients with B lymphoma associated with immunosuppressive therapy when the immunosuppressive regimen could be stopped or decreased.

Other adverse consequences of immunosuppression

Autoimmune diseases, such as rheumatoid arthritis, dermatomyositis, endocrinopathy, Sjögren's syndrome, vasculitis or lupus erythematosus, are more frequent in patients with a primary defect of the humoral immunity or the complement system. Similarly, immuno-allergic reactions have been suggested to be more frequent in immunocompromised patients. Severe hypersentivity reactions to medicinal products, such as sulphonamides, have been repeatedly described in patients with AIDS. Although an alteration in enzyme activities has been suggested or evidenced in some instances, it is unclear whether this mechanism can account for all the reported hypersensitivity reactions (Koopmans *et al.*, 1995). At the present time, no experimental or clinical data indicate that chemically-induced immunosuppression is associated with more frequent immuno-allergic reactions.

It has been long recognised that potent immunosuppressive agents, such as cyclo-phosphamide or cyclosporine, can enhance immune responses, in particular delayed-type hypersensitivity, in given conditions of drug administration in relation to antigen injection. These experimental findings which account for the use of the term 'immunomodulation' have no clear clinical relevance. Similarly, chemicals, such as cadmium, lead or selenium, were shown to enhance the immune response at very low levels of exposure, although higher levels of exposure were associated with immunosuppression (Zelikoff and Thomas, 1998). The dose–response curve is therefore quite unusual and the possibility remains that immunostimulation and immunosuppression can occur with the same chemical depending on the conditions of exposure. The possible health consequences of these findings remain to be established.

Immunosuppression and immunodepression

The majority of authors use the term 'immunosuppression' to describe a quantitative decrease in the immune response, regardless of the magnitude. This may, however, be misleading. Strictly speaking, immunosuppression is a profound abrogation (suppression) of the immune response, whereas immunodepression refers to a decrease in the immune response. In the latter situation, the immune response is only partly impaired. This distinction was stressed by others when using the term 'overimmunosuppression' to qualify the status of patients treated with potent immunosuppressive drugs.

From the perspective of immunotoxicity risk assessment, this distinction can be useful as the majority of treatments with medicinal products and exposures to environmental chemicals when found to exert direct immunotoxic effects, only depress the immune response and do not suppress it totally. One major issue is the concept of functional reserve capacity of the immune system: it is well recognised that redundant effector and/ or regulatory mechanisms are generally combined to mount a normal immune response. Therefore, when one such mechanism is impaired in relation to immunotoxic exposure, one or several mechanism(s) can exert their compensatory influence, so that the resulting consequence is the lack of overt change. Although antimicrobial resistance was consistently shown to be markedly impaired in patients treated with potent immunosuppressive agents, it is more difficult to identify whether and to what extent antimicrobial resistance is impaired following exposure to chemicals which depress, but do not totally abrogate, the immune response, that is to say immunodepressive chemicals.

Available human data, although limited, nevertheless suggest that chemically-induced immunodepression can be associated with more frequent infections. Supportive evidence comes from patients with autoimmune diseases, such as rheumatoid patients treated with low-dose immunosuppressive treatments. Opportunistic infections have been described, even though uncommonly in rheumatoid patients (Segal and Sneller, 1997). Similarly, human beings inadvertently exposed to brominated or chlorinated biphenyls, which are known to impair the immune response, were shown to develop more frequent respiratory infections (Kuratsune *et al.*, 1996). Very few data suggest that lymphomas can be more frequent in patients after chemical 'immunodepressive' exposures. However, lymphomas have been reported in a few rheumatoid patients treated with low-dose cyclophosphamide or methotrexate (Kamel *et al.*, 1995).

Although a limited body of evidence is available to support the hypothesis that mild-to-moderate impairment of the immune response (immunodepression) is a condition associated with more frequent infections and lymphomas, the issue whether susceptible individuals are at a greater risk of developing these complications should be raised. One example is the recent epidemic which killed thousands of seals from the North Sea (Harwood, 1989). Seals were infected by a mutant virus of the distemper virus family, but surprisingly the animals were far more susceptible to the virus than expected. The North Sea is heavily polluted by metals and hydrocarbons which are potently immuno-depressive in laboratory animals. When comparing seals fed contaminated herring from the North Sea to seals fed herring from the far less polluted Atlantic Ocean, statistically significant differences were noted in several key immune parameters.

In any case, the issue of whether chemical immunodepression can exacerbate a latent immune deficiency can be expanded to groups of the population that are likely to be more susceptible, such as infants, elderly people, and patients with AIDS. Although no firm answer is available, the coming years should bring helpful information and allow a more accurate risk assessment of immunotoxicants. In this context, the expected development of human (epidemiological) immunotoxicology is likely to be critical.

References

BIRKELAND, S.A., HAMILTON-DUTOIT, S., SANDVEJ, K., ANDERSEN, H.M.K., BENDTZEN, K., MØLLER, B. and JØRGENSEN, K.A. (1995) EBV-induced post-transplant lymphoproliferative disorder. *Transplant. Proceed.*, **27**, 3467–3472.

BODEY, G.P. and FAINSTEIN, V. (1986) Infections of the gastrointestinal tract in the immunocompromised patient. *Annu. Rev. Med.*, **37**, 271–281.

DAVIES, D.M. (1981) Effect of drugs on infections, In: *Textbook of Adverse Drug Reactions*, 2nd edition (Davies, D.M., ed.) pp. 518–527. Oxford: Oxford Press.

DONNELLY, J.P. (1995) Bacterial complications of transplantation: diagnosis and treatment. *J. Antimicrob. Chemother.*, **36**, Suppl. B, 59–72.

EHRKE, M.J., MIHICH, E., BERD, D. and MASTRANGELO, M.J. (1989) Effects of anticancer drugs on the immune system in humans. *Semin. Oncol.*, **16**, 230–253.

FRATTINI, J.E. and TRULOCK, E.P. (1993) Respiratory infections in immunocompromised patients. *Immunol. Allergy Clin. N. Am.*, **13**, 193–204.

GRIFFITHS, P.D. (1995) Viral complications after transplantation. *J. Antimicrob. Chemother.*, **36**, Suppl. B, 91–106.

HARWOOD, J. (1989) Lessons from the seal epidemic. *New Scientist*, 18 February, 38–42.

KAMEL, O.K., VAN DE RIJN, M., HANASONO, M.M. and WARNKE, R.A. (1995) Immunosuppression associated lymphoproliferative disorders in rheumatic patients. *Leuk. Lymph.*, **16**, 363–368.

KASS, E.H. and FINLAND, M. (1953) Adrenocortical hormones in infection and immunity. *Annu. Rev. Microbiol.*, **7**, 361–388.

KINLEN, L.J., SHEIL, A.G.R., PETO, J. and DOLL, R. (1980) Collaborative United Kingdom–Australasian study of cancer in patients treated with immunosuppressive drugs. *Br. Med. J.*, **2**, 1461–1466.

KOOPMANS, P.P., VAN DER VEN, A.J.A.M., VREE, T.B. and VAN DER MEER, J.W.M. (1995) Pathogenesis of hypersensitivity reactions to drugs in patients with HIV infection: allergic or toxic? *AIDS*, **9**, 217–222.

KURATSUNE, M., YOSHIMURA, H., HORI, Y., OKUMURA, M. and MASUDA, Y. (1996) *Yusho. A human disaster caused by PCBs and related compounds*. Fukuoka: Kyushu University Press.

LI, P.K.T., NICHOLLS, M.G. and LAI, K.N. (1990) The complications of newer transplant antirejection drugs: treatment with cyclosporin A, OKT3, and FK506. *Adv. Drug React. Acute Poisoning Rev.*, **9**, 123–155.

MACKAY, A., MACLEOD, T., ALCORN, M.J., LAIDLAW, M., MACLEOD, I.M., MILLAR, J.S., *et al.* (1980) Fatal disseminated BCG infection in an 18-year-old boy. *Lancet*, **ii**, 1332–1334.

MARGOLICK, J.B. and VOGT, R.F. (1991) Environmental effects on the human immune system and the risk of cancer: facts and fears in the era of AIDS. *Environ. Carcino. Ecotox. Rev.*, **9**, 155–206.

NICHOLSON, V. and JOHNSON, P.C. (1994) Infectious complications in solid organ transplant recipients. *Surg. Clin. N. Am.*, **74**, 1223–1244.

PENN, I. (1994) Malignancies. *Surg. Clin. N. Am.*, **74**, 1247–1257.

PILLAI, R., NAIR, B.S. and WATSON, R.R. (1991) AIDS, drugs of abuse and the immune system: a complex immunotoxicological network. *Arch. Toxicol.*, **65**, 609–617.

REVILLARD, J.P. (1990) Iatrogenic immunodeficiencies. *Curr. Op. Immunol.*, **2**, 445–450.

ROSEN, F.S., COOPER, M.D. and WEDGWOOD, R.J.P. (1995) The primary immunodeficiencies. *N. Engl. J. Med.*, **333**, 431–440.

RUBIN, R.H. (1993) Impact of cytomegalovirus infection on organ transplant recipients. *Rev. Infect. Dis.*, **12**, S754–S766.

RUBIN, R.H. and YOUNG, L.S. (1994) *Clinical Approach to Infection in the Compromised Host*, 3rd edition. New York: Plenum Medical.

SEGAL, B.H. and SNELLER, M.C. (1997) Infectious complications of immunosuppressive therapy in patients with rheumatic diseases. *Rheum. Dis. Clin. N. Am.*, **23**, 219–237.

SILVA, J., KAUFFMAN, C.A., SIMON, D.G., LANDRIGAN, P.J., HUMPHREY, H.E.B., HEATH, C.W., *et al.* (1979) Lymphocyte function in humans exposed to polybrominated biphenyls. *J. Reticuloendoth. Soc.*, **26**, 341–346.

SINGH, N. and YU, V.L. (1996) Infections in organ transplant recipients. *Curr. Op. Infect. Dis.*, **9**, 223–229.

SMIALOWICZ, R.J. (1996) The immunotoxicity of organic solvents. In: *Experimental Immunotoxicology* (Smialowicz, R.J. and Holsapple, M.P., eds) pp. 307–329. Boca Raton: CRC Press.

THOMAS, L. (1952) The effects of cortisone and adrenocorticotropic hormone on infection. *Annu. Rev. Med.*, **3**, 1–24.

VIAL, T., NICOLAS, B. and DESCOTES, J. (1996) Clinical immunotoxicology of pesticides. *J. Toxicol. Environ. Health*, **48**, 215–229.

ZELIKOFF, J.T. and THOMAS, P.T. (1998) *Immunotoxicology of Environmental and Occupational Metals.* London: Taylor & Francis.

4

Immunostimulation

Although the term 'immunomodulation' has long been preferred to immunostimulation, the introduction of recombinant cytokines into the clinical setting demonstrated that immunological responses as any other physiological responses, can indeed be increased (or stimulated). Other terms, such as immunorestoration or immuno-enhancement, have also been commonly used, but these linguistic differences are more likely to reflect limitations in our current understanding of the potential for pharmacological immuno-manipulation or immunostimulation than a scientifically established situation.

Overall, the adverse consequences of immunostimulation on human health have been less extensively investigated than those of immunosuppression, because immunosuppression has been an issue of primary concern for immunotoxicologists during the past 20 years, but also because potent immunostimulating agents, such as the recombinant cytokines interferon-α and IL-2, were introduced into the clinical setting only recently. Even though early medicinal products proposed for use as immunostimulants, such as levamisole, *Corynebacterium parvum* and thymic hormones, did not prove to exert marked therapeutic activity (Smalley and Oldham, 1984), adverse effects in some patients treated with these medicinal products were described more than ten years ago (Descotes, 1985a; Davies, 1986), and the clinical experience gained with the use of the most recent recombinant cytokines to treat human patients, largely confirmed these early reports.

Because the focus of immunotoxicologists has so often been on immunosuppression, very few medicinal products, and occupational or environmental chemicals have been recognised to exert unexpected or adverse immunostimulating properties. To some extent, the anti-H_2 histamine receptor antagonist cimetidine (Ershler *et al.*, 1983), the angiotensin-converting enzyme inhibitor captopril (Tarkowsky *et al.*, 1990), and the pesticide hexachlorobenzene (Michielsen *et al.*, 1997) are possible examples of xenobiotics with immunostimulating properties, even though the actual consequences on human health of the immunostimulation presumably associated with exposure to these compounds remain to be established.

Flu-like reactions

Hyperthermic reactions (>38–38.5°C) with chills, arthralgias and malaise were described very early following the introduction of medicinal products with suspected or established immunostimulating properties, such as glycan derivatives, *Corynebacterium parvum*, BCG, levamisole or poorly purified interferon preparations, into the clinical setting (Descotes, 1985a). Similar adverse reactions are typically noted following the administration of vaccines, particularly recall injections in small children. Such hyperthermic or flu-like reactions are nevertheless inconsistently seen and in most patients, they tend to be mild to moderate, and easily prevented or controlled with the administration of minor analgesic/ antipyretic agents, such as paracetamol (acetaminophen).

More recently, similar, but more severe, sometimes treatment-limiting, reactions were reported with recombinant interferon-α, IL-2, tumour necrosis factor (TNF-α) as well as therapeutical monoclonal antibodies and immunotoxins (Vial and Descotes, 1992, 1994, 1995a, 1995b, 1996). In these patients, hyperthermia can reach 40°C or even more, and is associated with severe symptoms, such as diarrhoea, vomiting, chest pain, hypotension possibly leading to cardiovascular collapse or cardiac ischaemia, and neurological disorders, such as tremor, confusion, obnubilation and seizures. Slower administration, lower dose and treatment with oral pentoxyphylline or indomethacin were shown to reduce the incidence and severity of this syndrome. The terms 'acute cytokine syndrome' or 'cytokine release syndrome' tend to be used instead of flu-like reaction to describe such severe adverse effects.

The mechanism of flu-like reactions is not fully understood (Vial and Descotes, 1995a). It no longer appears to be related to a direct effect of interferons or IL-2, but rather to be the consequence of the acute release of fever-promoting factors in the hypothalamus, such as eicosanoids, IL-1 and TNF-α. Interestingly, such adverse reactions have been noted with immunostimulating substances which can activate macrophages, either directly (such as microbial extracts) or via the release of interferon-γ, which in turn results in the release TNF-α and IL-1. *In vitro* assays have been designed to measure the TNF-α and/or IL-1 releasing properties of therapeutic cytokines, monoclonal antibodies or biotechnology-derived products with a reasonably good predictibility for clinical adverse reactions in humans (Robinet *et al.*, 1995).

The release of IL-1, IL-6 and TNF-α was also noted following the inhalation of metal fumes, in particular zinc fumes, suggesting that acute fume reactions in welders might involve an immunostimulating mechanism, presumably via macrophage activation (Blanc *et al.*, 1991).

Exacerbation of underlying diseases

Shortly after their introduction into the clinical setting, immunostimulating agents were found to exacerbate or facilitate a variety of underlying diseases in treated patients, namely dormant diseases, autoimmune diseases and immuno-allergic reactions. Early and speculative reports were later largely confirmed following the introduction of therapeutic recombinant cytokines to treat patients with a variety of pathological conditions.

Dormant diseases

When the early immunostimulating agents were introduced into the clinical setting, they were often empirically administered to patients with varied and supposedly or conclusively documented immunopathological conditions. For instance, patients given levamisole developed clinical symptoms attributed to the exacerbation of the treated disease, such as immune complex vasculitis, rheumatoid arthritis, Crohn's disease, and chronic brucellosis. Similar, although less frequent, adverse reactions were reported with other immunostimulating agents, such as thymic hormones, BCG, or microbial extracts (Descotes, 1985a).

More recent data confirmed early findings: IL-2 and/or interferon-α treatments were reported to exacerbate or reactivate psoriasis, lichen planus, sarcoidosis, or chronic hepatitis (Vial and Descotes, 1996). As discussed later, initiation of treatment with immunostimulating agents can also be associated with exacerbation of immune-allergic reactions induced by unrelated allergens.

Autoimmune diseases

The introduction of therapeutic recombinant cytokines into the clinical setting showed that immunostimulation can be associated with more frequent autoimmune diseases (Miossec, 1997; Vial and Descotes, 1995a).

An unexpectedly large number of patients treated with IL-2 and/or interferon-α have been reported to present with varied autoimmune diseases, such as autoimmune thyroiditis, lupus erythematosus, insulin-dependent diabetes mellitus, myasthenia gravis, autoimmune haemolytic anaemia, polymyositis, autoimmune hepatitis or Sjögren's syndrome (Vial and Descotes, 1992, 1994, 1995b). Thyroid disorders are by far the most common findings. Interferon treatments were associated with thyroid diseases, including hypothyroidism, hyperthyroidism and biphasic thyroiditis, in 5 to 12 per cent of patients and the majority of these patients had serum antithyroid auto-antibodies, i.e. thyroid microsomal and antithyroglobulin auto-antibodies. Thyroid disease is still more common in patients treated with IL-2 (up to 35 per cent). Most of these patients present with hypothyroidism and antithyroid auto-antibodies. Thyroid disorders have also been described in a few patients treated with interferon-γ or GM-CSF.

Overall, the fact that autoimmune diseases are more frequent in patients treated with recombinant cytokines provides further evidence, if need be, for the pivotal role of cytokines in the pathogenesis of many (immuno)pathological conditions. Because of their pleiotropic effects and redundancy, it is unlikely that one single cytokine can be the key determinant in the occurrence of immune disorders or immune-mediated clinical symptoms, but it may act as a booster (Vial and Descotes, 1995a). Among the many hypotheses proposed to account for the observed clinical and immunological changes in patients treated with recombinant cytokines, an abnormal expression of MHC class II molecules induced by interferon-γ, and amplified by IL-1 and TNF-α, is the most commonly accepted hypothesis. Thus, under the influence of interferon-γ, thyroid cells would express MHC class II molecules and act as antigen-presenting cells with the production of antithyroid auto-antibodies as a consequence.

Importantly, the clinical and biological signs of autoimmune diseases associated with immunostimulating agents are similar to those of the corresponding spontaneous diseases, in sharp contrast to most drug-induced autoimmune reactions, as discussed later in this volume (see Chapter 6).

Autoimmune reactions have been reported following various chemical exposures, but the mechanism involved is unknown and no evidence is currently available to suggest that immunostimulation might be the causative mechanism. In addition, the actual incidence of autoimmune diseases associated with occupational or environmental chemical exposures is not known. Some authors claimed that chemical exposures are common causative factors, but so far no conclusive epidemiological data have confirmed these claims.

Immuno-allergic reactions

Another potential adverse consequence of immunostimulation is the facilitation or induction of immune-allergic reactions to unrelated allergens. Surprisingly, this issue has very seldom been raised after the introduction of the first immunostimulating agents, even though it seemed logical to assume that immunostimulating agents could stimulate normal as well as 'abnormal' immune responses.

Tilorone, an interferon inducer developed in the early 1970s, was shown to increase IgE production in the rat, whereas total serum IgE levels and skin eruptions were found to be more frequent in rheumatoid patients treated with levamisole than in non-treated rheumatoid controls (Descotes, 1985a). Data from the post-marketing surveillance of immunostimulating agents are scarce, but exacerbation of asthma, eczema and rhinitis has been reported shortly after the introduction of these medicinal products in some patients.

Interestingly, immune-allergic reactions have been described in patients treated with recombinant cytokines (Vial and Descotes, 1992, 1994, 1995b). These reactions include tubulo-interstitial nephritis, angioneurotic oedema, and linear IgA bullous dermatosis. Finally, cancer patients treated with IL-2 were shown to develop significantly more hypersensitivity reactions to radiological contrast media than matched cancer patients without a history of IL-2 treatment, but these findings failed to be confirmed by other investigators (Vial and Descotes, 1995a).

Inhibition of hepatic drug metabolism

It has long been recognised that infectious diseases, particularly viral infections (Koren and Greenwald, 1985; Kraemer *et al.*, 1982; Renton and Knickle, 1990), can depress cytochrome P450-dependent biotransformation pathways. Further to pioneer findings with interferon inducers, the majority of early medicinal products with immunostimulating properties, including interferons α and γ, bacterial extracts such as BCG, *Corynebacterium parvum* and *Bordetella pertussis*, and bacterial or viral vaccines, have been shown to inhibit hepatic drug metabolism, either *in vitro* or in laboratory animals (Descotes, 1985b; Renton, 1983).

Fewer human data are however available, but BCG and interferon-α were found to influence negatively the pharmacokinetics of theophylline, antipyrine and/or caffeine in healthy volunteers as well as in patients (Craig *et al.*, 1993; Gray *et al.*, 1983; Williams and Farrell, 1986; Jonkman *et al.*, 1989), whereas conflicting results were observed with influenza and tetanus vaccines (Grabenstein, 1990).

Several cytokines, including interferon-α (IFN-α), IFN-β, IFN-γ, IL-1, IL-2, IL-6 and TNF-α, have been shown to depress hepatic microsomal cytochrome P450-mediated metabolism after *in vitro* or *in vivo* administration to rats and/or mice (Renton and Armstrong, 1994). IL-1 is likely to play a pivotal role. Pathological conditions associated with hyperproduction of IL-1 have long been known to result in an inhibition of cytochrome P450-dependent pathways. Such pathological conditions include viral infections in humans and adjuvant-induced arthritis in rats. IL-2 was also shown to potentiate the effects of phenobarbital. That TNF-α was shown to depress hepatic drug metabolism *in vivo*, but not *in vitro*, suggests that IL-1 acts as a mediator of TNF-α effects. More recent experimental work suggested that IL-6 is actually the key factor. IL-6 production is increased by IL-1 and hepatocytes have IL-6 receptors. IL-6 was shown to downregulate CYP1A1,

CYP1A2, CYP1A3 and CYP2B *in vitro*, but little or no effect was evidenced after *in vivo* administration.

The molecular mechanism of cytochrome P450 downregulation by immunostimulating agents is relatively well understood. In contrast to other known inhibitors of cytochrome P450, such as cimetidine and erythromycin, which bind to and inhibit several cytochrome P450 isoenzymes, neither IL-1 or IL-6 have been shown to bind to cytochrome P450. Interestingly, immunostimulating agents which can effectively increase interferon production downregulate cytochrome P450-dependent pathways of hepatic drug metabolism. Interferons are known to inhibit protein synthesis and downregulation of cytochrome P450 has been repeatedly suggested to involve alterations in apoprotein synthesis or breakdown, and not a general decrease in the synthesis of hepatic microsomal proteins. However, immunostimulating agents acting through non-interferon-mediated effects have also been shown to downregulate several cytochrome P450 isoforms. A decrease in several cytochome P450 mRNAs has been evidenced and the possibility raised that a common, but still putative, intermediate, which is unlikely to be IL-6, is involved.

Immunostimulation and immunosuppression

Immunosuppressive drugs, such as cyclophosphamide or cyclosporine, have been shown to enhance antigen-specific immune responses depending on the dose and time of administration in relation to antigen injection. Based on these early experimental works, the term 'immunomodulating agents' was coined and once preferred to immunostimulants, although no clinical benefit was actually found to be obtained.

In contrast, medicinal products with immunostimulating properties can sometimes prove to be immunosuppressive (Vial and Descotes, 1996). In sharp contrast to early expectations, immunostimulating drugs have so far been largely unsuccessful in curing of most malignancies. Early findings suggested that tumour facilitation could even occur. Concern was also raised regarding BCG immunotherapy, interferon-α2A, and IL-2 alone or associated with interferon-β. Malignancies, such as acute leukaemia and Kaposi's sarcoma, have been described in a few patients. In addition, an increased incidence of clinically significant infectious complications affecting the urinary tract or the catheter site, has been repeatedly reported in patients treated with IL-2 infusion. Interestingly, impaired humoral and cell-mediated immune responses have been found in patients treated with high dose IL-2.

References

BLANC, P., WONG, H., BERNSTEIN, M.S. and BOUSHEY, H.A. (1991) An experimental human model of metal fume fever. *Ann. Intern. Med.*, **114**, 930–936.

CRAIG, P.I., TAPNER, M. and FARRELL, G.C. (1993) Interferon suppresses erythromycin metabolism in rats and human subjects. *Hepatology*, **17**, 230–235.

DAVIES, I. (1986) Immunological adjuvants of natural origin and their adverse effects. *Adv. Drug React. Ac. Pois. Rev.*, **1**, 1–21.

DESCOTES, J. (1985a) Adverse consequences of chemical immunomodulation. *Clin. Res. Pract. Drug Regul. Affairs*, **3**, 45–52.

DESCOTES, J. (1985b) Immunomodulating agents and hepatic drug-metabolizing enzymes. *Drug Metab. Rev.*, **16**, 175–184.

ERSHLER, W.B., HACKER, M.P., BURROUGHS, B.J., MOORE, A.L. and MYERS, C.F. (1983) Cimetidine and the immune response. I. In vivo augmentation of nonspecific and specific immune responses. *Clin. Immunol. Immunopathol.*, **26**, 10–17.

GRABENSTEIN, J.D. (1990) Drug interactions involving immunologic agents. Part I. Vaccine-vaccine, vaccine-immunoglobulin, and vaccine drug interactions. *DICP, Ann. Pharmacother.*, **24**, 67–81.

GRAY, J.D., RENTON, K.W. and HUNG, O.R. (1983) Depression of theophylline elimination following BCG vaccination. *Br. J. Clin. Pharmac.*, **16**, 735–737.

JONKMAN, J.H.G., NICHOLSON, K.G., FARROW, P.R., *et al.* (1989) Effects of interferon-α on theophylline pharmacokinetics and metabolism. *Br. J. Clin. Pharmac.*, **27**, 795–802.

KOREN, G. and GREENWALD, M. (1985) Decrease in theophylline clearance causing toxicity during viral epidemics. *J. Asthma*, **22**, 75–79.

KRAEMER, M.J., FURUKAWA, C., KOUP, J.P. and SHAPIRO, G. (1982) Altered theophylline clearance during an influenza outbreak. *Pediatrics*, **69**, 476–480.

MICHIELSEN, C.P.P.C., BLOKSMA, N., ULTEE, A., VAN MIL, F. and VOS, J.G. (1997) Hexachlorobenzene induced immunomodulation and skin and lung lesions: a comparison between Brown Norway, Lewis and Wistar rats. *Toxicol. Appl. Pharmacol.*, **144**, 12–26.

MIOSSEC, P. (1997) Cytokine-induced autoimmune disorders. *Drug Safety*, **17**, 93–104.

RENTON, K.W. (1983) Relationships between the enzymes of detoxication and host defense mechanisms. In: *Biological Basis of Detoxication* (Caldwell, J. and Jacovy, W.B., eds), pp. 387–324. New York: Academic Press.

RENTON, K.W. and ARMSTRONG, S.G. (1994) Immune-mediated downregulation of cytochrome P450 and related drug biotransformation. In: *Immunotoxicology and Immunopharmacology*, 2nd edition (Dean, J.H., Luster, M.I., Munson, A.E. and Kimber, I., eds), pp. 501–512. New York: Raven Press.

RENTON, K.W. and KNICKLE, L.C. (1990) Regulation of cytochrome P450 during infectious disease. *Can. J. Physiol. Pharmacol.*, **68**, 777–781.

ROBINET, E., MOREL P., ASSOSSOU, O. and REVILLARD, J.P. (1995) TNF-alpha production as an in vitro assay predictive of cytokine-mediated toxic reactions induced by monoclonal antibodies. *Toxicology*, **100**, 213–223.

SMALLEY, R.V. and OLDHAM, R.K. (1984) Biological response modifiers: preclinical evaluation and clinical activity. *CRC Crit. Rev. Oncol. Hematol.*, **1**, 259–280.

TARKOWSKY, A., CARLSTEN, H., HERLITZ, H. and WESTBERG, G. (1990) Differential effects of captopril and enalapril, two angiotensin converting enzyme inhibitors, on immune reactivity in experimental lupus disease. *Agents Actions*, **31**, 96–101.

VIAL, T. and DESCOTES, J. (1992) Clinical toxicity of interleukin-2. *Drug Safety*, **7**, 417–433.

VIAL, T. and DESCOTES, J. (1994) Clinical toxicity of the interferons. *Drug Safety*, **10**, 115–150.

VIAL, T. and DESCOTES, J. (1995a) Immune-mediated side-effects of cytokines in humans. *Toxicology*, **105**, 31–57.

VIAL, T. and DESCOTES, J. (1995b) Clinical toxicity of cytokines used as hemopoietic growth factors. *Drug Safety*, **13**, 371–406.

VIAL, T. and DESCOTES, J. (1996) Drugs acting on the immune system. In: *Meyler's Side Effects of Drugs*, 13th edition (Dukes, M.N.G., ed.), pp. 1090–1165. Amsterdam: Elsevier Sciences.

WILLIAMS, T. and FARRELL, G.C. (1986) Inhibition of antipyrine metabolism by interferon. *Br. J. Clin. Pharmacol.*, **22**, 610–612.

5

Hypersensitivity

Hypersensitivity reactions due to medicinal products and industrial as well as environmental chemicals are relatively common and often claimed to be increasingly more common. They are considered to account for the majority of adverse reactions involving the immune system in humans (Bernstein, 1997; DeShazo and Kemp, 1997).

Although limited reliable information is available, hypersensitivity reactions have been proposed to account for at least 20 to 30 per cent of all recorded adverse drug reactions based on the experience of post-marketing drug surveillance. Hypersensitivity reactions, mainly contact dermatitis and respiratory allergy, are the most common complaints at the workplace (Kimber and Dearman, 1997; Kimber and Maurer, 1996). Finally, allergies or sensitivities in relation to environmental exposures have been claimed to affect a significant fraction of the general population, although a reliable diagnosis is very difficult to obtain in most cases of multiple chemical sensitivity syndrome (Miller, 1996).

A major problem when dealing with hypersensitivity reactions induced by xenobiotics is the current lack of clear understanding of the mechanism or most certainly mechanisms involved in so many instances. A number of reactions have been considered to be immune-mediated (that is to say the consequence of immunological response), but in fact not all are unequivocally antigen-specific (DeSwarte, 1986; Wedner, 1987). Other reactions, possibly the most common, could involve non-immune mediated mechanisms, namely pharmacological or toxicological mechanisms (Doenicke, 1980). Importantly, limited progress has been achieved in recent years to ascertain the mechanisms which are actually involved despite the impressive advances gained in our understanding of the immune system and immune responses. Because the mechanism is so seldom established or attainable, few reliable diagnostic tools are available and their value is often a matter of debate (Rieder, 1997). In addition, the clinical symptoms seen in patients who develop hypersensitivity reactions involving presumably widely different mechanisms, are often confusingly similar, leading to erroneous conclusions, even though the incriminated compound is a prototypic allergen, such as penicillin G (Sturtees et al., 1991).

Probably due to all these uncertainties, the term 'allergy' is often used undiscriminatingly to describe these reactions. 'Hypersensitivity' or 'hyperergia', which suggest an abnormal susceptibility of the affected patient, whatever the mechanism involved, are suggested to be more appropriate terminology. Use of the term 'allergy', or better 'immuno-allergy', should be restricted to those reactions that are presumably or clearly caused by an antigen-specific immune response. When clinical symptoms mimic an immuno-allergic reaction, but the mechanism is unlikely to be immune-mediated, the term 'pseudo-allergy' should be preferred. Finally, the term 'idiosyncrasy' is proposed to characterise only those non-immune mediated reactions which are thought to involve a pharmacogenetic predisposition (Pirmohamed *et al.*, 1997), in order to avoid confusion between pseudo-allergic and idiosyncratic reactions.

The scope of this chapter is restricted to immuno-allergic and pseudo-allergic reactions.

Immunogenicity

By definition, immuno-allergic reactions involve antigen-specific immunological responses mediated by either specific antibodies and/or sensitised lymphocytes (DeShazo and Kemp, 1997; DeSwarte, 1986; Wedner, 1987). The most characteristic feature of immuno-allergic reactions due to xenobiotics is therefore the pivotal role of highly specific immunological recognition mechanisms, so that both the sensitisation to and the immunogenicity of xenobiotics are the absolute hallmarks of immuno-allergic reactions. Strictly speaking, the diagnosis of immuno-allergic reactions should not be envisaged unless the involvement of specific immunological mechanisms can be documented at least to some extent.

Xenobiotics as haptens

A prior contact with the antigen is an absolute prerequisite for the development of an immuno-allergic reaction. As it is usually impossible to obtain conclusive evidence whether a prior contact with a given xenobiotic in a given patient was actually sensitising, the assumption has to be made that the prior contact was sensitising, hence a first level of uncertainty in the diagnosis of immuno-allergic reactions induced by xenobiotics. After a subsequent (but not necessarily the second) contact, sensitisation becomes patent, as a clinical reaction develops.

To be sensitising, xenobiotics must comply with both the following prerequisites:

- Xenobiotics must be foreign or 'non-self' to be sensitising, which is always so when xenobiotics are concerned, except for a limited number of medicinal products of human origin, such as biotechnologically-derived human insulin or growth hormone.

- The molecular weight of xenobiotics must be large enough for them to be sensitising (or immunogenic), although no minimal requirement is established (but presumably above 5000 D or less for peptides because of the greater immunogenicity of peptidic bonds).

Foreign macromolecules, proteins, polypeptides or microbial extracts can thus be directly sensitising or immunogenic. The molecular weight of most xenobiotics, particularly medicinal products, except for a very few, such as heparin or insulin, is far too small (usually below 500 D), so that they cannot be directly immunogenic, but must act as haptens after strongly binding to carrier macromolecules to become indirectly immunogenic. Importantly, typical drug binding to plasma proteins, such as albumin, is not strong (energy-rich) enough to result in the formation of immunogenic complexes so that no relationship exits between the degree of binding to plasma albumin and immunogenicity (as an illustrative example, diazepam, of which over 95 per cent is bound to plasma proteins, is an extremely rare cause of immune-mediated hypersensitivity reactions).

Because binding must be much stronger for a hapten–carrier immunogenic complex to be formed *in vivo*, the chemical reactivity of xenobiotics must be high. In contrast to many industrial products, which have sufficient chemical reactivity, medicinal products have limited, if no, chemical reactivity to avoid undue toxic effects, such as genotoxicity or teratogenicity. This should be kept in mind when extrapolating results obtained in animal models with industrial respiratory and contact skin sensitisers to the far less chemically reactive medicinal products. A widely held assumption is that metabolites instead of the corresponding parent molecule are actually involved in hapten formation (Hess and Rieder, 1997; Park *et al.*, 1987; Shapiro and Shear, 1996). As biotransformation processes can lead to very reactive, intermediate metabolites, it is tempting to believe that metabolites actually play the role of haptens. This mechanism was conclusively shown to be involved in penicillin allergy, as penicillin by-products, instead of penicillin itself, have been shown to bind covalently to macromolecules resulting in the production of anti-hapten antibodies, such as anti-penicilloyl antibodies and, to a lesser extent, anti-penaldate, anti-penalmadate, or anti-penicillenate antibodies, which can be detected in the sera of patients with a history of immuno-allergic reaction to penicillin (Ahlstedt and Kristofferson, 1982). Direct evidence that this mechanism is involved in the majority of, if not all, cases is lacking even though this is ignored by many authors. One explanation is the very short half-life of highly reactive intermediates, which are very quickly metabolised into more stable metabolites, and are therefore extremely difficult to detect and identify. Nevertheless, the vast majority of available data, even though usually indirect, argues for the role of metabolites as haptens.

If the role of intermediate metabolites in hapten formation is accepted, the immune system must be expected to discriminate between extremely close chemical structures, such as metabolites and the corresponding parent molecule, so that *in vitro* tests currently proposed for the diagnosis of drug and chemical allergy should presumably generate many false negative results since they are obtained using inappropriate probes, that is to say the parent molecules, instead of the supposedly causative intermediate metabolites. This is another cause of uncertainty in the diagnosis of immuno-allergic reactions to xenobiotics, particularly medicinal products, but surprisingly, more emphasis has so far been on false positive than false negative results of *in vitro* allergy testing (DeShazo and Kemp, 1997; Rieder, 1997).

Risk factors

Whatever the immunogenic (or sensitising) potential of a given xenobiotic, not all individuals obviously become sensitised upon exposure, so that the role of contributing factors should be considered (Hoigné *et al.*, 1983). For instance, less than 1–2 per cent of treated patients are expected to develop an immune-mediated reaction to penicillin G and this is an unusually high incidence of such reactions associated with drug treatment.

Suggested or confirmed contributing factors are the following (Hoigné *et al.*, 1983; Vervloet *et al.*, 1995):

- *Age*: young adults generally develop more frequent immune-allergic reactions to medicinal products and chemicals than young children and elderly people for reasons that remain unclear, but immuno-allergic reactions in young children and elderly people are usually more severe.
- *Gender*: women are slightly more affected than men, but the role of gender does not seem to be essential. It is unclear whether the hormonal status accounts for this slight

difference. A complex interaction between age and sex is possible, as suggested by experimental studies: for instance, contact sensitisers, such as dinitrochlorobenzene (DNCB), induced greater contact skin responses in young female than in old male mice.

- *Atopy*: conflicting results have been published regarding the predisposing role of atopy. This may be explained by the inconsistent definition of this condition (hay fever, allergic rhinitis to common allergens, reaginic asthma or constitutional dermatitis). It seems however that personal atopy should not be considered a major, if at all, predisposing factor to immune-mediated adverse drug reactions.

- *Route of exposure*: although every route of administration or exposure is potentially sensitising, certain routes result in more frequent sensitisation than others. Sensitisation is much more frequent after topical application, whereas the oral route is generally associated with the induction of tolerance, as shown experimentally. However, the mechanisms resulting in the break of oral tolerance are unknown. When the parenteral route is used (intravenous route for instance), reactions tend to be more severe.

- *Exposure regimen*: intermittent exposures markedly facilitate sensitisation and the development of immuno-allergic reactions, as shown by intermittent treatments with rifampicin and chloramphenicol, which, for instance, were reported to induce much more frequent immune-mediated tubulo-interstitial nephritis and aplastic anaemia, respectively (Léry *et al.*, 1978; Poole *et al.*, 1971). In disagreement with the general belief, a small dose of one hapten may not be sufficient to trigger an immune-allergic reaction in previously sensitised hosts. A threshold dose is likely to exist beyond which immune-allergic reactions can develop, but it is also possible that different threshholds can be identified between individuals.

- *Individual (genetic) predisposition*: all individuals are not equal regarding the risk of sensitisation to medicinal products and xenobiotics. In the same conditions of penicillin treatment, only a fraction of patients develop an immuno-allergic reaction. A genetic predisposition was suggested to be involved, but early findings that HLA determinants, such as HLA-DR4 and HLA-DR6, could play a role, failed to be confirmed or further documented. The search for genes predisposing to allergy is nevertheless still underway (Marsh and Meyers, 1992).

- *Biotransformation/pharmacokinetics*: the role of metabolic or pharmacokinetic traits, involving obvious possible genetic relations, should also be considered. As intermediate by-products, instead of the parent molecule, are more likely to be involved in drug sensitisation, biotransformation pathways should be considered as essential (Shapiro and Shear, 1996).

- *Chemical structure*: this is another critical, but ill-understood factor of chemical (and drug as well) immunogenicity (Baldo and Harle, 1990; Baldo and Pham, 1994; Basketter *et al.*, 1995) despite current extensive efforts to search for components of the chemical structure which are more likely to be involved in immunogenicity. More information is obviously needed before it can be possible to identify which components of the chemical structure are specifically involved in the sensitising potential of xenobiotics.

Immuno-allergic reactions

The mechanisms of immuno-allergic reactions to xenobiotics are still ill-understood today. Gell and Coombs proposed a pathogenic classification of four types (Coombs and

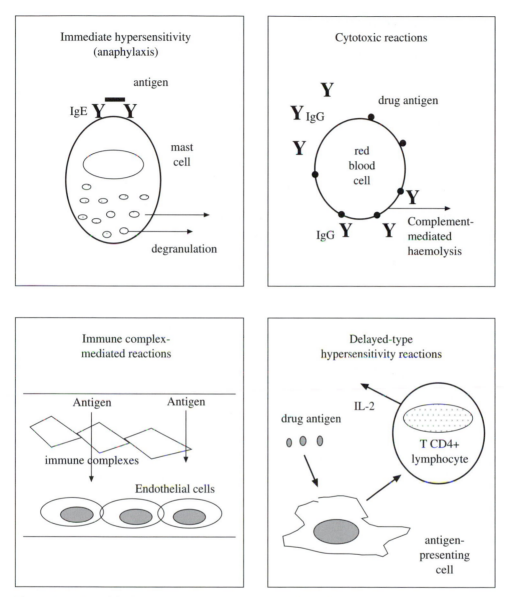

Figure 5.1 Simplified representation of immune-mediated hypersensitivity reactions to drugs according to Coombs and Gell classification

Gell, 1968), which is still often referred to, even though it is actually not helpful to understand every aspect of drug- and chemical-induced immune-allergic reactions. Firstly, immuno-allergic reactions are quite artificially dissected into separate and supposedly exclusive mechanisms in spite of the fact that several mechanisms are likely to co-exist in the same patient. Secondly, in contrast to antibody-mediated mechanisms, which were given much emphasis in the Coombs and Gell classification, cellular mechanisms are pivotal, as suggested by the most recent advances in our understanding of the immune response. Therefore, the Coombs and Gell classification seems much more adapted to the teaching of undergraduate students than to the diagnosis and investigation of drug-induced immune-mediated hypersensitivity reactions.

Immediate hypersensitivity or anaphylaxis

Anaphylaxis is an immediate-type hypersensitivity reaction involving specific IgE anti-bodies (Atkinson and Kaliner, 1992; DeJarnatt and Grant, 1992). Following a (supposed or documented) sensitising contact, specific IgE antibodies bind to high affinity receptors (FcgRI) on the membrane surface of target cells, namely mast cells in tissues, and basophils in the peripheral blood. After a subsequent, but not necessarily the next contact, the reaction between a divalent antigen and specific IgE antibodies bound on target cells, induces the degranulation of the target cells with the immediate release of stored vasoactive mediators (e.g. histamine), and sets off the synthesis of eicosanoid derivatives, such as prostaglandins and leukotrienes, through the release of arachidonic acid controlled by phospholipase A2. These mediators, which are either stored or subsequently synthesised, exert a variety of biological effects responsible for the acute clinical signs of anaphylaxis, namely urticaria, angioedema, bronchospasm and shock. However, drug-induced anaphylaxis is a relatively rare, but life-threatening event (Van der Klauw *et al.*, 1996).

   The biological diagnosis of anaphylactic reactions is based on the detection of specific IgE (Rieder, 1997). Total IgE levels are valueless, as they do not demonstrate that specific IgE actually account for possibly increased IgE levels. Specific IgE are best detected in the sera of patients using the radioallergosorbent test (RAST), but due to limited availability, *in vitro* assays which mimic the *in vivo* IgE-dependent degranulation of basophils are to be preferred. The human basophil degranulation test has never been the matter of sound and comprehensive validation, so that claims of its uselessness are certainly not based on grounds more scientifically conclusive than claims of its reliability. Cytosolic granules in which basophils store histamine, are specifically stained by toluidine blue, and therefore can easily be counted with a microscope. Following controlled incuba-tion of basophil suspensions with graded concentrations of the suspected xenobiotic, a second basophil count is performed and the basophil degranulation test is considered positive when at least 30 per cent of basophils have degranulated. Unfortunately, this human basophil degranulation test is time-consuming and poorly standardisable. The histamine release assay is nowadays the preferred assay. Following incubation of periph-eral blood cells with increasing concentrations of the suspected xenobiotic, histamine is assayed in the supernatant, preferably using ELISA or radioimmunoassay (Van Toorenbergen and Vermeulen, 1990). More recently, the diagnostic value of skin tests was repeatedly stressed (Bryunzeel and Maibach, 1997). Readings are performed after four hours and, in experienced hands, they do not seem to induce anaphylactic complica-tions, as seen or feared in the past, particularly when low concentrations or patch tests are used. Cellular assays, such as the lymphocyte transformation test and the lymphocyte migration inhibition test, are largely but inconsistently considered obsolete (Nyfeler and Pichler, 1997). There are indeed reasons to believe that the initial concept is still valid: that cytokines are released upon antigen-specific activation of T lymphocytes cannot be denied. *In vitro* cell assays taking advantage of recent technical advances to measure cytokine levels (e.g. ELISA) or cytokine mRNAs are worth further consideration (Bienvenu *et al.*, 1998).

Cytotoxic reactions

IgM and more rarely IgG antibodies are involved in cytotoxic reactions. Typically, these are acute adverse haematological reactions caused either by the sensitising chemical bound to the surface of blood cells (white or red cells, platelets), which conflict with specific circulating antibodies and activate complement resulting in cytolysis, or by antibodies bound to blood cells, or to circulating immune complexes (Habibi, 1988).

Clinically, cytotoxic reactions manifest as immuno-allergic haemolytic anaemias, thrombocytopenias or agranulocytosis. The diagnosis in humans can be facilitated when the phenomenon is reproduced *in vitro*, which unfortunately is not often possible or available.

Immune complex reactions

In the presence of antigens in greater quantity in the serum than IgM or IgG antibodies, circulating immune complexes are formed, which deposit in capillary vessels and activate the complement system and neutrophils. Lesions develop in endothelium cells and result in localised or generalised vasculitis. In this latter case, the typical clinical pattern is serum sickness, with fever, arthralgias, cutaneous eruption and proteinuria, nine to 11 days after the injection of heterologous serum, such as antithymocyte globulin (Cunningham *et al.*, 1987). Cutaneous vasculitis, a localised immune-complex mediated reaction, is associated with a typical histological pattern of leukocytoclasia, which is often considered as conclusive and sufficient evidence of drug-induced vasculitis.

Recently, cases of serum sickness disease have been described following the administration of varied medicinal products, with the cephalosporin cefaclor as one of the most common causes (Vial *et al.*, 1992). The clinical features are similar to those of a typical serum sickness, despite the lack of immune complex-mediated renal injury or even the detection of immune complexes in the sera of patients. As no circulating or deposited immune complexes are detected, the term 'serum sickness-like disease' is often used to describe such conditions.

Delayed hypersensitivity

Typically, delayed hypersensitivity includes allergic contact dermatitis and photoallergy, which can be reproduced in humans with the use of cutaneous or photocutaneous tests.

Allergic contact dermatitis

Contact dermatitis can be either non-immunologically mediated irritant (irritant contact dermatitis) or cell-mediated (allergic contact dermatitis). Both conditions are sometimes clinically inseparable. Irritant contact dermatitis, which is much more frequent (80 per cent of cases), may be complicated by a superimposed allergic contact dermatitis.

Allergic contact dermatitis is a delayed-type hypersensitivity reaction characterised by the infiltration of T lymphocytes into the dermis and epidermis (Krasteva, 1993). It is caused by skin contact with a chemical which triggers an immunological response leading to inflammatory skin lesions. A hapten for contact hypersensitivity is usually a low-molecular-weight, lipid-soluble substance which binds or complexes with cell surface or structural proteins on various cells, including Langerhans cells and keratinocytes. Langerhans cells process and present the antigen to T lymphocytes, which leads to the clonal proliferation of sensitised lymphocytes and to a clinically patent inflammatory reaction. Acute allergic contact dermatitis usually develops as an erythematous, vesicular, oedematous eruption at the sites of skin contact with the sensitising substance. Following repeated exposure, chronic contact dermatitis will develop with erythematous, scalling and thickened skin lesions. Although contact dermatitis can be either irritant or allergic, these conditions may be difficult to differentiate.

Medicinal products are common causes of allergic contact dermatitis (Storrs, 1991). Major contact sensitisers among medicinal products include antibiotics, such as neomycin and penicillins, local anaesthetics, transdermal drug delivery systems, and pharmaceutical preservatives, such as merthiolate. Systemic drug-induced contact dermatitis may also occur after oral, intramuscular, or intravenous administration. Occupational and environmental chemicals are also common causes of allergic contact dermatitis (Stewart, 1992; Suskind, 1990).

Photosensitivity

Photosensitivity is far less common (Maurer, 1983; Allen, 1993). In fact, most photosensitivity reactions induced by drugs and chemicals are phototoxic. Phototoxicity is generally characterised as exaggerated sunburn of rapid onset after sun exposure. Phototoxic drugs and chemicals absorb energy from UVA light and release it into the skin, causing cellular damage. Photoallergy is an immune-mediated reaction. Light can induce structural changes in a drug so that it acts as a hapten and binds to skin proteins. Photo-allergy occurs rarely compared with phototoxicity.

Pseudo-allergic reactions

Although immuno-allergic reactions induced by xenobiotics are considered to be relatively common, an immune-mediated mechanism is unlikely to be consistently involved. The term 'pseudo-allergic' reactions was coined when it became obvious that adverse clinical manifestations similar to those of immune-allergic reactions could be observed despite the documented lack of any previous (and supposedly sensitising) contact (Doenicke, 1980). Clinical similarities are related to the involvement of the same vasoactive mediators which, however, are released by non-immunological mechanisms in pseudo-allergic reactions.

Non-pseudo-allergic reactions

Discrepancies can be found in the use of the term 'pseudo-allergic reaction'. For the sake of clarity, it seems better to exclude patients with a pharmacogenetic defect predisposing to rare adverse reactions. The term 'idiosyncrasy' is then recommended. A very good, historical, example is the development of acute haemolysis in patients with a congenital deficiency in the enzyme glucose-6-phosphatase dehydrogenase, following treatment with non-steroidal anti-inflammatory or antimalaria agents.

A number of adverse reactions have been considered to be pseudo-allergic, even though the same vasoactive, proinflammatory mediators as those involved in immune-mediated reactions are definitely not involved.

Hoigné's syndrome

Hoigné's syndrome, actually described for the first time by Batchelor in 1951, is characterised by an acute feeling of thoracic tightness and malaise, with hypotension following the injection of the slow-release formulation procaine-penicillin (Hoigné and Schock, 1956). The likeliest mechanism is the formation of procaine microcristal aggregates resulting in pulmonary micro-emboli. With an incidence of 1 out of 1000 injections, this syndrome was often erroneously considered as an allergic reaction to penicillin; it is likely that a majority of elderly people today are considered allergic to penicillin because they presented with Hoigné's syndrome much earlier in their life as they were treated with procaine-penicillin to prevent the cardiac complications of rheumatic fever.

Local anaesthetic-induced shock

Local anaesthetics can cause acute adverse reactions. The ester derivatives, such as procaine and tetracaine, are potent sensitisers, whereas the amide derivatives, such lignocaine and mepivacaine, are not. True immune-allergic systemic reactions to local anaesthetics are considered to be extremely rare (Gall *et al.*, 1996). Shock associated with loco-regional anaesthesia is more likely to be either a vagal reaction induced by pain and fear, or a systemic reaction due to the inadvertent intra-arterial injection of the local anaesthetic. The offending role of drug preservatives, such as parabens, has been documented in a few illustrative patients.

Ampicillin skin rash

Ampicillin and its derivatives frequently cause mild cutaneous eruptions (Shapiro *et al.*, 1969). Several findings argue against an immune-allergic mechanism:

- the unusual time course of events (onset between the second and fourth day of treatment);
- the lack of clinical symptoms suggestive of anaphylaxis;
- the usual negativity of allergological tests; and
- the greater incidence of viral infections (particularly infectious mononucleosis and cytomegalovirus infections).

Nevertheless, ampicillin and its derivatives can also induce anaphylactic reactions, hence the absolute need for a careful chronological and semiological analysis of any skin eruption associated with these antibiotics to avoid potentially harmful rechallenge in previously sensitised patients.

Non-specific histamine release

Histamine, which is stored in mast cells and basophils, can be released independently of any IgE-mediated mechanism. A cytotoxic or osmotic effect may be involved, if not fully substantiated. Histamine-releasing compounds include morphine and the morphine derivatives codeine and meperidine, and general intravenous anaesthetics. However, it is difficult to ascertain the clinical consequence of a compound with histamine-releasing properties demonstrated in human volunteers. The measurement of serum histamine is seldom conclusive due to the very short half-life of histamine, so that the role of direct histamine release in anaphylactoid reactions associated with general anaesthesia, although frequently claimed in the past (Doenicke, 1980), remains speculative. Surprisingly, although direct histamine release is often considered a proven mechanism of acute reactions induced by general anaesthesia, a majority of studies found a lack of histamine release following injection of intravenous anaesthetics.

However, the fact that direct histamine release can lead to adverse clinical reaction is best illustrated by morphine and morphine derivatives, as pretreatment with anti-H_1 and anti-H_2 antihistamines was shown to diminish the incidence and/or severity of adverse reactions associated with histamine release, particularly in the course of thoracic and open-heart surgery (Doenicke and Lorenz, 1982). Finally, intravenous administration of the antimicrobial vancomycin is associated with acute clinical reactions involving documented histamine release, commonly called the 'red man syndrome' (Wallace *et al.*, 1991).

Direct activation of the complement system

Complement is a complex system, the activation of which is under the control of several regulatory proteins along two different pathways, the classical and alternate pathways. The activation of complement by either pathway can be caused by immunological as well as non-immunological stimuli, among which Cremophor El and hydrosoluble radiological contrast media are the most commonly cited medicinal products.

Cremophor El is a pharmaceutical solvent used to dissolve poorly soluble drugs. Acute pseudo-allergic reactions due to Cremophor El have been recorded with intravenous formulations of diazepam, vitamin K1, alfadione and more recently, cyclosporin. A non-specific activation of complement was shown to be involved in acute reactions associated with administration of the general intravenous anaesthetic alfadione (Benoit *et al.*, 1983). Obviously, when Cremophor El is involved as in an intravenous formulation, no risk can be expected when using an oral formulation of the same drug without cremophor, as exemplified by cyclosporin (Liau-Chu *et al.*, 1997).

Hydrosoluble radiological contrast media induce quite similar adverse reactions (Bush and Swanson, 1991). Such reactions are relatively common (5 to 20 per cent of patients), but usually mild to moderate with only one death in approximately 40 000 radiological examinations. An immuno-allergic mechanism is very unlikely to be involved, despite the concept of 'iodine allergy' which is commonly held by the public. The role of complement activation, in particular the alternate pathway, has been suggested to be involved, but no conclusive evidence has been provided and other biological systems, such as fibrinolysis, the kinins and the coagulation cascade, are also likely to play a role.

Intolerance to non-steroidal anti-inflammatory agents

Non-steroidal anti-inflammatory drugs (NSAIDs), of which aspirin is the leading compound, can induce acute intolerance reactions, mainly precipitations of asthma attacks noted in 10 per cent of asthmatics (Manning and Stevenson, 1991; Szczelick, 1997). Intolerance reactions typically develop within one hour after aspirin ingestion as an acute asthma attack, often associated with rhinorrhoea and conjunctival irritation. A certain diagnosis is based on provocation tests only, either oral, inhaled or nasal tests. Urticaria with or without concomitant angioedema is less frequent in patients intolerant to aspirin.

One single individual can develop stereotyped reactions following ingestion of different NSAIDs. As chemical structures are too dissimilar, cross-allergenicity cannot be involved and an impact on the release of arachidonic derivatives is far more likely. NSAIDs inhibit the enzyme cyclo-oxygenase (COX), of which two isoforms, namely

COX-1 and COX-2, are known. Aspirin and the majority of NSAIDs are much more potent inhibitors of COX-1 than COX-2. Any NSAID with marked COX-1 inhibiting activity can precipitate asthma attacks in contrast to those NSAIDs, such as nimesulide, with preferentially anti-COX-2 activity. Importantly, COX inhibitors devoid of anti-inflammatory effects in humans, such as paracetamol, can nevertheless induce acute intolerance reactions.

There is no firm evidence that COX inhibition results in increased availability of arachidonic acid and/or leukotriene release. It is only possible that inhibition of COX is associated with the overproduction of cysteinyl-leukotrienes which are important mediators of asthma, as evidenced by the use of recently developed antileukotriene drugs. Why a given patient will develop aspirin intolerance remains elusive. The facilitating role of chronic inflammation, persistent viral infection, or genetic predisposition has been suggested.

Other compounds have been reported to induce somewhat similar clinical reactions in patients intolerant to aspirin and NSAIDs. These compounds with no known COX inhibiting activity include azo dyes, such as tartrazine, sulphites, or monosodium glutamate. In fact, it is at best unclear whether the same mechanism is involved.

References

AHLSTEDT, S. and KRISTOFFERSON, A. (1982) Immune mechanisms for induction of penicillin allergy. *Prog. Allergy*, **30**, 67–134.

ALLEN, J.E. (1993) Drug-induced photosensitivity. *Clin. Pharm.*, **12**, 580–587.

ATKINSON, T.P. and KALINER, M.A. (1992) Anaphylaxis. *Med. Clin. N. Am.*, **76**, 841–855.

BALDO, B.A. and HARLE, D.G. (1990) Drug allergenic determinants. *Monogr. Allergy*, **28**, 11–51.

BALDO, B.A. and PHAM, N.H. (1994) Structure-activity studies on drug-induced anaphylactic reactions. *Chem. Res. Toxicol.*, **7**, 703–721.

BASKETTER, D., DOOMS-GOOSENS, A., KARLBERG, A.-T. and LEPOITTEVIN, J.P. (1995) The chemistry of contact allergic: why is a molecule allergenic? *Contact Derm.*, **32**, 65–73.

BENOIT, Y., CHADENSON, O., DUCLOUX, B., VEYSSEYRE, C.M., BRINGUIER, J.P. and DESCOTES, J. (1983) Hypersensitivity to Althesin infusion: measurement of complement involvement. *Anaesthesia*, **38**, 1079–1081.

BERNSTEIN, D.I. (1997) Allergic reactions to workplace allergens. *JAMA*, **278**, 1907–1913.

BIENVENU, J., MONNERET, G., GUTOWSKI, M.C. and FABIEN, N. (1998) Cytokine assays in human sera and tissues. *Toxicology*, **129**, 55–61.

BUSH, W.H. and SWANSON, D.P. (1991) Acute reactions to intravascular contrast media; types, risk factors, and specific treatment. *Am. J. Radiol.*, **157**, 1153–1161.

BRYUNZEEL, D.P. and MAIBACH, H.I. (1997) Patch testing in systemic drug eruptions. *Clin. Dermatol.*, **15**, 479–484.

COOMBS, R.R.A. and GELL, P.G.H. (1968) Classification of allergic reactions responsible for drug hypersensitivity reactions. In: *Clinical Aspects of Immunology*, 2nd edition (Coombs, R.R.A. and Gells, P.G.H., eds), pp. 575–596. New York: F.A. Davis.

CUNNINGHAM, E., CHI, Y., BRENTJENS, J. and VENUTO, R. (1987) Acute serum sickness with glomerulonephritis induced by antithymocyte globulin. *Transplantation*, **43**, 309–312.

DEJARNATT, A.C. and GRANT, A.A. (1992) Basic mechanisms of anaphylaxis and anaphylactoid reactions. *Immunol. Allergy Clin. N. Am.*, **12**, 501–515.

DESHAZO, R.D. and KEMP, S.F. (1997) Allergic reactions to drugs and biologic agents. *JAMA*, **278**, 1895–1906.

DESWARTE, R.D. (1986) Drug allergy: an overview. *Clin. Rev. Allergy*, **4**, 443–469.

DOENICKE, A. (1980) Pseudo-allergic reactions due to histamine release during intravenous anaesthesia. In: *Pseudo-Allergic Reactions* (Dukor, P., Kallós, P., Schlumberger, H.D. and West, G.B., eds), pp. 224–250. Basel: Karger.

DOENICKE, A. and LORENZ, W. (1982) Histamine release in anaesthesia and surgery. Premedication with H1- and H2-receptors antagonists: indications, benefits and possible problems. *Klin. Wochenschr.*, **60**, 1039–1045.

GALL, H., KAUFMANN, R. and KALVERAM, C.M. (1996) Adverse reactions to local anaesthetics: analysis of 197 cases. *J Allergy Clin. Immunol.*, **97**, 933–937.

HABIBI, B. (1988) Drug-induced immune hemolytic anemias. *Path. Biol.*, **36**, 1237–1245.

HESS, D.A. and RIEDER, M.J. (1997) The role of reactive drug metabolites in immune-mediated adverse drug reactions. *Ann. Pharmacother.*, **31**, 1378–1387.

HOIGNÉ, R. and SCHOCK, K. (1956) Anaphylactic shock and acute non-allergic reaction after procaine penicillin. *Schweiz. Med. Wochenschr.*, **89**, 1350–1356.

HOIGNÉ, R., STOCKER, F. and MIDDLETON, P. (1983) Epidemiology of drug allergy. In: *Allergic Reactions of Drugs* (De Weck, A.L. and Bundgaard, H. eds), pp. 187–205. Handbook of Experimental Pharmacology, vol. 63. Berlin: Springer.

KIMBER, I. and DEARMAN, R.G. (1997) *Toxicology of Respiratory Hypersensitivity*. London: Taylor & Francis.

KIMBER, I. and MAURER, T. (1996) *Toxicology of Contact Hypersensitivity*. London: Taylor & Francis.

KRASTEVA, M. (1993) Contact dermatitis. *Int. J. Dermatol.*, **32**, 547–560.

LÉRY, N., DESCOTES, J. and EVREUX, J.Cl. (1978) A review of chloramphenicol-induced blood disorders. *Vet. Hum. Toxicol.*, **6**, 177–181.

LIAU-CHU, M., THEIS, J.G.W. and KOREN, G. (1997) Mechanism of anaphylactoid reactions: improper preparation of high-dose intravenous cyclosporine leads to bolus infusion of Cremophor El and cyclosporine. *Ann. Pharmacother.*, **31**, 1287–1291.

MANNING, M.E. and STEVENSON, D.D. (1991) Pseudo-allergic drug reactions. Aspirin, nonsteroidal antiinflammatory drugs, dyes, additives, and preservatives. *Immunol. Allergy Clin. N. Am.*, **11**, 659–678.

MARSH, D.G. and MEYERS, D.A. (1992) A major gene for allergy – fact or fancy? *Nature Genetics*, **2**, 252–254.

MAURER, T. (1983) *Contact and Photocontact Allergens*. New York: Marcel Dekker.

MILLER, C.S. (1996) Chemical sensitivity: symptom, syndrome or mechanism for disease? *Toxicology*, **111**, 69–86.

NYFELER, B. and PICHLER, W.J. (1997) The lymphocyte transformation test for the diagnosis of drug allergy: sensitivity and specificity. *Clin. Exp. Allergy*, **27**, 175–181.

PARK, B.K., COLEMAN, J.W. and KITTERRINGHAM, N.R. (1987) Drug disposition and drug hypersensitivity. *Biochem. Pharmacol.*, **36**, 581–590.

PIRMOHAMED, M., MADDEN, S. and PARK, B.K. (1997) Idiosyncratic drug reactions: metabolic activation as a pathogenic mechanism. *Clin. Pharmacokinet.*, **31**, 215–230.

POOLE, G., STRADLING, P. and WORLLEDGE, S. (1971) Potentially serious side effects of high-dose twice-weekly rifampicin. *Br. Med. J.*, **3**, 343–347.

RIEDER, M.J. (1997) In vivo and in vitro testing for adverse drug reactions. *Pediatr. Clin. N. Am.*, **44**, 93–111.

SHAPIRO, L.E. and SHEAR, N.H. (1996) Mechanisms of drug reactions: the metabolic track. *Semin. Cutan. Med. Surg.*, **15**, 217–227.

SHAPIRO, S., SLONE, D., SISKIND, V., *et al.* (1969) Drug rash with ampicillin and other penicillins. *Lancet*, **ii**, 969–972.

STEWART, L.A. (1992) Occupational contact dermatitis. *Immunol. Allergy Clin. N. Am.*, **12**, 831–846.

STORRS, F.J. (1991) Contact dermatitis caused by drugs. *Immunol. Allergy Clin. N. Am.*, **11**, 509–523.

STURTEES, S.J., STOCKTON, M.G. and GIETZEN, T.W. (1991) Allergy to penicillin: fable or fact? *Br. Med. J.*, **302**, 1051–1052.

SUSKIND, R.R. (1990) Environment and the skin. *Med. Clin. N. Am.*, **74**, 307–324.

SZCZELICK, A. (1997) Mechanism of aspirin-induced asthma. *Allergy*, **52**, 613–619.

VAN DER KLAUW, M.M., WILSON, J.H.P. and STRICKER, B.H.Ch. (1996) Drug-associated anaphylaxis: 20 years of reporting in The Netherlands (1974–1994) and review of the literature. *Clin. Exp. Allergy*, **26**, 1355–1363.

VAN TOORENBERGEN, A.W. and VERMEULEN, A.M. (1990) Histamine release from human peripheral blood leukocytes analyzed by histamine radioimmunoassay. *Agents Actions*, **30**, 278–280.

VERVLOET, D., PRADAL, M., CHARPIN, D. and PORRI, F. (1995) Diagnosis of drug allergic reactions. *Clin. Rev. Allergy Immunol.*, **13**, 265–280.

VIAL, T., PONT, J., PHAM, E., RABILLOUD, R. and DESCOTES, J. (1992) Cefaclor-associated serum sickness-like disease: eight cases and review of the literature. *Ann. Pharmacother.*, **26**, 910–914.

WALLACE, M.R., MASCOLA, J.R. and OLDFIELD, E.C. (1991) Red man syndrome: incidence, etiology, and prophylaxis. *J. Infect. Dis.*, **164**, 1180–1185.

WEDNER, H.J. (1987) Allergic reactions to drugs. *Primary Care*, **14**, 523–545.

6

Autoimmunity

Besides decreases or increases in the normal immune response, namely direct immunotoxicity and hypersensitivity, a third situation related to immunotoxicity is autoimmunity (Bigazzi, 1994; Kammüller *et al.*, 1989; Kilburn and Warshaw, 1994). Nowadays, autoimmunity is no longer considered to be basically a consequence of immune stimulation, as it once was considered to be, together with hypersensitivity. However, it is increasingly apparent that some common fundamental mechanisms can lead to immune responses potentially resulting either in hypersensitivity or in autoimmune reactions (Griem *et al.*, 1998).

The diagnosis of autoimmune reactions induced by medicinal products and occupational or environmental chemicals is very seldom convincingly documented. Nevertheless, autoimmunity is often claimed to be a major health issue despite our current lack of clear understanding of the mechanim(s) involved and of the actual incidence and clinical severity of most autoimmune diseases. It is in fact unsure, or unlikely, that autoimmune reactions induced by medicinal products and chemicals are as common as often claimed (Vial *et al.*, 1994).

The presence of auto-antibodies in the sera of patients is an absolute prerequisite for the diagnosis of autoimmune diseases, and more particularly chemically-induced autoimmune reactions (Peter and Shoenfeld, 1996). Typically, two types of chemically-induced autoimmune reactions, namely organ-specific and systemic autoimmune reactions, with quite distinctive features, can be differentiated.

Organ-specific autoimmune reactions

Organ-specific autoimmune reactions induced by xenobiotics are characterised by a homogeneous antibody response against a unique target (resulting in the presence of predominant types of auto-antibody in the sera of affected patients) and by clinical symptoms closely mimicking those found in the corresponding spontaneous autoimmune disease.

Autoimmune haemolytic anaemia

In addition to a direct toxic effect, two immune-mediated mechanisms can result in haemolytic anaemias: specific cytotoxic antibodies involved in immuno-allergic haemolytic anaemias and and autoimmunity (Holland and Spivak, 1992).

Autoimmune haemolytic anaemias are rare and heterogeneous. The majority of auto-immune haemolytic anaemias are spontaneous or secondary to neoplasia, but in a few instances, they are caused by drug treatments, primarily α-methyldopa (Perry *et al.*, 1971). Up to 30 per cent of patients treated with the antihypertensive drug α-methyldopa have a positive Coombs test and auto-antibodies against Rhesus erythrocyte antigens. However, haemolysis is patent, as mild anaemia with reticulocytosis, non-conjugated hyperbilirubinaemia, and diminished serum haptoglobulin, in less than 1 per cent of patients. In very rare patients, anaemia is severe and potentially fatal. Anaemia usually recovers in the few days following the cessation of the offending drug, but serum auto-antibodies can still be found after a few months. Importantly, no antibodies are directed against α-methyldopa or one of its identified metabolites. Essentially, auto-antibodies are directed against the Rhesus antigens and it is usually impossible to differentiate auto-antibodies in drug-induced and spontaneous autoimmune haemolytic anaemias. Establishing a causal relationship is therefore not easy in most cases, but the time course of events and the recovery after drug treatment cessation are helpful. Even though the mechanism involved in the production of these auto-antibodies is unknown, their role in the induction of haemolysis is well established.

A small number of medicinal products, such as chlorpropamide, L-dopa, fludarabine, mefenamic acid, nomifensine and procainamide, have been reported to induce autoimmune haemolytic anaemias, which nevertheless remain a very uncommon drug-induced adverse effect. No occupational or environmental chemicals have so far been suspected to induce autoimmune haemolytic anaemias, but underreporting must be considered as a critically contributing factor.

Drug-induced myasthenia

Myasthenia is an autoimmune disease characterised by a loss of muscular strength due to auto-antibodies against the nicotinic receptors of the neuromediator acetylcholine, located in the neuromuscular motor plates, resulting in inhibition of the binding of the para-sympathetic mediator and of the transmission of the nerve influx to the muscles (Massey, 1997). Virtually all patients have ocular symptoms, most frequently ptosis and diplopia. Weakness of the oropharyngeal muscles may produce difficulties for chewing, swallowing, speaking or breathing.

Penicillamine is the main cause of drug-induced myasthenias (Wittbrodt, 1997). The clinical features are generally close to those of spontaneous myasthenia with ptosis and diplopia as the initial clinical symptoms. Auto-antibodies to acetylcholine receptors are detected in the sera of approximately 75 per cent of patients. Despite the lack of correlation between the duration of treatment and/or the dose used, the causal relationship is established from the slow decrease of clinical symptoms and auto-antibodies, which is never observed in spontaneously occurring myasthenia. Other pharmaceutical products which can induce myasthenia, include the antirheumatics tiopronine and pyritinol, and the antiepileptic drug trimethadione.

Table 6.1 Medicinal products most frequently involved in organ-specific autoimmune reactions

Organ-specific autoimmune reactions	Main causative xenobiotics
Haemolytic anaemia	α-Methyldopa
	L-dopa
	Chlorpropamide
	Fludarabine
	Mefenamic acid
	Nomifensine
	Procainamide
Myasthenia	Penicillamine
	Tiopronin
	Pyritinol
	Trimethadione
Pemphigus	Captopril
	Penicillamine
	Pyritinol
	Tiopronine

Pemphigus

Pemphigus is a relatively rare skin disease, which is characterised by bullous skin eruptions of varying severity. Histologically, pemphigus is characterised by destruction of the epidermis (acantholysis) and IgG auto-antibodies against the intercellular substance in the deep epidermis. A few drugs can induce pemphigus, namely captopril, and the anti-rheumatics penicillamine, pyritinol and tiopronine (Anhalt, 1989).

Polymyositis

Polymyositis is an uncommon disease, very rarely induced by medicinal products. It is characterised by muscular deficiency, myalgias and sometimes cutaneous lesions. It is more frequent in rheumatoid patients treated with the antirheumatic penicillamine (Halla *et al.*, 1984).

Immunotoxic ('autoimmune') hepatitis

A few drugs, such as the diuretic tienilic acid, the antihypertensive dihydralazine, the volatile general anaesthetic halothane and the antidepressant iproniazid, can cause hepatitis associated with the presence of highly specific auto-antibodies in the sera of affected patients (Beaune *et al.*, 1994). Strictly speaking, these are not autoimmune hepatitis, hence the preferred term 'immunotoxic', as a mechanism closer to sensitisation than autoimmunity has been shown to be involved. However, the ambiguity of the terminology reflects both ill-understood possible similarities and recognised differences in the mechanisms involved (Griem *et al.*, 1998). Induction of the immune response is related to drug-induced structural changes in specific hepatocyte constituents following biotransformation of the offending drug into metabolites with the capacity to bind to hepatocyte constituents.

The LKM2 antibody in tienilic acid-induced hepatitis is one of the best examples (Beaune *et al.*, 1987). Over 500 cases of hepatic injury, of which at least 25 were fatal, were reported in the 1980s. Tienilic acid is biotransformed into a reactive metabolite by cytochrome P450 2C9. Auto-antibodies, which react with liver and kidney sections from untreated rats (LKM2 auto-antibodies), have been found in the sera of patients with tienilic acid-induced hepatitis, but not in the sera of patients with non-tienilic acid-induced hepatitis or in the sera of non-hepatitic patients treated with tienilic acid. LKM2 auto-antibodies were found to recognise cytochrome P450 2C9 in humans, the cytochrome P450 isoform involved in the metabolism of tienilic acid, suggesting that a reactive metabolite of tienilic acid can bind to the producing cytochrome P450 2C9 and induce an immune-mediated destruction of this cytochrome.

Similarly, dihydralazine is biotransformed by cytochrome P450 1A2 into reactive radicals, which covalently bind to cytochrome P450. Anti-liver microsome auto-antibodies have been found in patients with dihydralazine-induced hepatitis (Bourdi *et al.*, 1990).

Mechanism(s) of organ-specific drug-induced autoimmune reactions

The mechanism of organ-specific drug-induced autoimmune reactions is unknown. Various hypotheses have been proposed. In addition, it is very unlikely that one single mechanism can account for such a variety of reactions.

Immunopharmacological mechanism

One early hypothesis was the depression of suppressor T cell functions (when suppressor T cells were considered pivotal components of the immunological orchestra) resulting in the polyclonal activation of B lymphocytes and the abnormal production of auto-antibodies (Kirtland *et al.*, 1980). However, decreases as well as increases in suppressor T cell activity were evidenced in patients with autoimmune haemolytic anaemia induced by α-methyldopa. As suppressor T-cells are no longer considered to exist, this hypothesis can hardly be sustained. However, it could be interesting to investigate whether the cytokine profile of patients with α-methyldopa induced autoimmune haemolytic anaemia is more Th_1 or Th_2-like, which seemingly has so far not been considered.

Penicillamine exerts a variety of immunopharmacological properties, the exact relevance of which is at best debatable with regard to the therapeutic activity of this compound, and it is still impossible to explain why it can induce organ-specific autoimmune reactions in treated patients. From the available knowledge, the immunopharmacological hypothesis cannot account for the reported drug-induced autoimmune reactions, and in particular the specificity of auto-antibodies.

Alteration of cellular constituents

That alteration of cellular constituents by the offending drug could result in the formation of neo-autoantigens and auto-antibodies, is unlikely, but could not be formally ruled out using α-methyldopa or an oxidised metabolite of α-methyldopa (Gootlieb and Wurzel, 1974). However, the auto-antibodies detected in the sera of patients treated with α-methyldopa were not different from those found in patients with spontaneous autoimmune haemolytic anaemia and were never found to be directed against α-methyldopa or one of its metabolites.

The role of thiol groups in organ-specific autoimmune reactions induced by penicillamine and several medicinal products, such as tiopronine and captopril, has been suggested, but failed to be formally substantiated (Pfeiffer and Irons, 1985). Experimentally, penicillamine was shown to bind to subunits of the acetylcholine receptor by forming disulphide bridges, which could result in the production of antireceptor auto-antibodies. However, these auto-antibodies did not cross-react with penicillamine and were identical to those detected in the sera of patients with spontaneous myasthenia. Therefore, the thiol group hypothesis, even though attractive, remains to be conclusively documented and can be considered as a likely oversimplification of much more complex immune and non-immune mechanisms.

Another mechanism to be considered is the proposed involvement of reactive metabolites generated by activated leukocytes, possibly resulting in autoimmune, but also hypersensitivity and idiosyncratic reactions (Uetrecht, 1994).

Systemic autoimmune reactions

Systemic autoimmune reactions induced by xenobiotics are characterised by a heterogeneous antibody response directed against ubiquitous cell targets (resulting in the presence of varied auto-antibodies in the sera of affected patients) and by a pattern of clinical symptoms hardly resembling that found in the corresponding spontaneous autoimmune disease. Systemic autoimmune reactions, even though uncommon, can be caused by a variety of medicinal products and a few occupational and possibly environmental chemical exposures (Bigazzi, 1994).

Lupus syndrome

The lupus syndrome or pseudolupus is the most frequent drug-induced autoimmune reaction, even though it is extremely rare (Adams and Hess, 1991; Price and Venables, 1995).

Clinically, drug-induced lupus syndromes are very dissimilar to lupus erythematosus. The most typical clinical signs of drug-induced lupus syndromes include arthritis, fever, weight loss, and muscular weakness. Cutaneous manifestations are often limited to uncharacteristic skin eruptions, such as erythema. Renal signs are inconsistent and mild when present. No neurological signs are noted in contrast to the neurolupus found in severe lupus erythematosus. One major distinctive sign is the occurrence of pleural and/ or pericardial effusions, which are relatively common and potentially severe in drug-induced lupus, but seldom seen in patients with systemic lupus erythematosus. No biological signs are typical of a drug-induced lupus syndrome. Auto-antibodies are IgG in most instances. High antihistone antibody titres are found in 95 per cent of patients with the lupus syndrome. Interestingly, serum antiDNA antibodies are sDNA antibodies (namely native DNA antibodies) in patients with the lupus syndrome and dDNA antibodies (namely denatured DNA antibodies) in patients with lupus erythematosus, with few exceptions. The lupus syndrome used to have a favourable outcome after cessation of the offending drug. The majority of clinical signs disappear within a few weeks, except for the pleural/ pericardial effusions which may persist for several months.

Hydralazine and procainamide are the main causes of drug-induced lupus syndromes (Adams and Hess, 1991; Price and Venables, 1995). Auto-antibodies have been detected in the sera of up to 25 per cent of patients treated with hydralazine and 50 per cent of

Table 6.2   Major drug and toxic causes of the lupus syndrome

Most commonly involved drugs	
Hydralazine	Procainamide
Other involved drugs	
Allopurinol	Lithium
α-Methyldopa	Lovastatin
β-Blockers (practolol, acebutolol, etc.)	Minocycline
Captopril	Nitrofurantoin
Carbamazepine	Penicillamine
Chlorpromazine	Phenylbutazone
Dapsone	Propylthiouracil
Diphenylhydantoin	Quinidine
Ethosuximide	Sulphonamides
Isoniazid	Trimethadione
Occupational/environmental exposures	
Chlordane	Silicosis
Chlorpyrifos	Thallium
Formaldehyde	Trichloroethylene
Hydrazine	

patients treated with procainamide. A relation was found between the prevalence of auto-antibodies, the dose and/or duration of treatment, and the slow acetylator phenotype. However, no clinical signs were associated with auto-antibodies in the majority of patients.

The other causes of drug-induced lupus syndromes were more seldom noted. They include several antiepileptic drugs, such as trimethadione, diphenylhydantoin and carbamazepine, beta-blockers, in particular acebutolol, the major tranquiliser chlorpromazine and the anti-tuberculous drug isoniazid. Lupus syndromes induced by industrial and environmental chemicals have been exceptionally described: thallium, hydrazine, hair dyes, and trichloroethylene, have been suggested to be involved. Auto-antibodies have been detected in the sera of patients with silicosis. The possible relation between systemic lupus erythematosus and environmental pollution has also been advocated, but rarely, if ever, documented.

Scleroderma-like diseases

Scleroderma is characterised by a more or less diffuse infiltration of the dermis by collagen, the synthesis of which by fibroblasts is increased. Scleroderma can be localised to a few zones of the dermis, or be generalised with systemic complications involving the gut, the broncho-pulmonary tract, the heart and the kidneys. The production of IL-1 and various growth factors under the control of T lymphocytes and/or the monocytes/macrophages seems to be the primary event leading to scleroderma. An autoimmune phenomenon is increasingly considered to be involved.

Few drugs and chemicals have been reported to induce scleroderma-like diseases (Bourgeois and Aeschlimann, 1991). The most severe was the so-called oculomucocutaneous syndrome induced by the beta-blocker practolol. This syndrome included keratoconjunctivitis, lesions of the conjunctivae with loss of sight, psoriasis-like eruption, and pleural and/or pericardial effusion. The drug was withdrawn in 1975. Workers with silicosis have been shown to be at a greater risk of developing scleroderma and the association silicosis–scleroderma has been described as Erasmus syndrome. Occupational exposures to vinyl chloride or trichloroethylene were also suggested to be associated with scleroderma-like diseases. Finally, a major debate in recent years was the possible link of scleroderma with the implantation of silicon prostheses, particularly for breast replacement after surgical removal for cancer, or for breast enlargement (Wong, 1996). Many epidemiological studies were conducted to demonstrate or rule out a causal link, and even though most studies failed to provide convincing evidence for a causal relationship, the possibility remains that epidemiological studies are not sensitive enough to detect slight increases in the incidence of a pathological condition associated with toxic exposure.

Mechanisms

The mechanisms of systemic autoimmune reactions are not known despite extensive research (Kammüller *et al.*, 1989). A careful analysis of clinical symptoms associated with systemic drug-induced autoimmune reactions tends to wonder why these reactions are still considered as the iatrogenic counterparts of spontaneous autoimmune diseases. In fact, clinical symptoms often diverge and biological disturbances are conflicting. Various pathophysiological mechanisms have been proposed, but none has so far been accepted or conclusively shown to be involved.

Activation of dormant autoimmune abnormalities

In sharp contrast to autoimmune findings associated with immunostimulation, the activation of dormant autoimmune abnormalities, or a latent viral infection, is no longer accepted as an attractive hypothesis to account for drug-induced autoimmune reactions (Adams and Hess, 1991). Activation of dormant autoimmune abnormalities was also often suggested to be involved in supposedly more frequent rheumatoid arthritis associated with oestroprogestative contraceptives. But this assumption is no longer held as valid or clinically relevant (Pladevall-Vila *et al.*, 1996).

Interactions with nuclear macromolecules

Interactions with nuclear macromolecules, as formerly proposed for the commonly involved medicinal products hydralazine and procainamide, is now considered unlikely to be involved. Hydralazine and procainamide can indeed bind to DNA and nucleoproteins (Yamauchi *et al.*, 1975), but the concentrations required are much too high to support a specific binding. In addition, the auto-antibodies detected in the sera of patients with the drug-induced lupus syndrome never reacted with the incriminated drug or one of its recognised metabolites. No one could actually reproduce the human disease in laboratory animals after the binding of hydralazine or procainamide to various nuclear macromolecules. Similarly, the role of anti-practolol antibodies could never be confirmed to be, as initially suggested (Amos *et al.*, 1978), the causative mechanism of the oculomucocutaneous syndrome induced by the beta-blocker practolol.

Immunopharmacological mechanism

The role of immunomodulation is unlikely to be involved in drug-induced systemic autoimmune reactions as the majority of causative drugs have no clearly established immunopharmacological effects (Adams and Hess, 1991). A deficiency in complement components can facilitate autoimmune diseases; for instance, decreased C4 levels or a hypofunctional C4 component have been found to be associated with impaired clearance of immune complexes and more frequent autoimmune diseases. As a number of drugs can modulate the complement system, it is tempting to propose that drug treatment might result from pharmacological interference with the complement system. Available results with penicillamine, hydralazine and procainamide are largely contradictory or inconclusive, and this assumption cannot be based on documented data.

Risk factors

When analysing the pathogenetic mechanisms of systemic autoimmune reactions, the role of risk factors should be considered carefully (Adams and Hess, 1991). Hormones are likely to play a role, as oestrogens facilitate the development of autoimmune diseases in laboratory animals, and presumably account for the marked female predominance of spontaneous as well as chemically-induced autoimmune diseases.

A pharmacogenetic predisposition is another risk factor to be considered, as patients with the slow acetylator phenotype have been shown to develop more lupus syndromes when treated with long-term hydralazine and procainamide. However, several inbred strains of rodents which have been shown to develop experimental diseases more frequently, still develop the disease inconsistently, which suggests that additional factors are required. That toxic factors could be involved remains largely unsettled (Kardestuncer and Frumkin, 1997).

Pseudo-GvH diseases

An interesting hypothesis regarding the mechanism of drug-induced systemic autoimmune diseases is the concept of pseudo-GvH diseases.

The initial hypothesis

The administration of histo-incompatible lymphocytes to a host can induce a potentially lethal, graft-versus-host (GvH) reaction (Ferrara and Degg, 1991). The clinical symptoms observed after bone marrow graft include hyperpyrexia, cutaneous eruptions, arthralgias, lung infiltrates and adenopathies. Remarkably, these clinical symptoms are similar to those observed in patients with drug-induced systemic autoimmune reactions, such as the lupus syndrome or scleroderma-like disease. Thanks to the early work of Helga and Ernst Gleichmann (Gleichmann *et al.*, 1984), the hypothesis of a pseudo-GvH reaction came to light. Mice were injected into one footpad with the antiepileptic drug diphenylhydantoin, the adverse effects of which mimic a GvH reaction. Seven days later, a significant increase in popliteal lymph node weight was evidenced. Interestingly, a regional GvH reaction was known to be revealed by the injection of lymphocytes from one parent into the footpad of an F_1 rat bred from parents derived from different syngenic strains, which induced an increased weight of the popliteal lymph node. This was the starting point of the development of the popliteal lymph node assay (see Chapter 16).

The mechanism of pseudo-GvH reactions is not clearly understood at the present time. An interference with MHC class II molecules is the most attractive hypothesis. In many instances, specific immune responses require that the antigen is presented to $CD4^+$ T lymphocytes by antigen-presenting cells, which process the antigen before presentation as small peptides on their surface. After antigen recognition within the context of the MHC class II restriction, CD4+ T lymphocytes are activated, then release IL-2 which in turn activates other lymphocytes. The overactivation of CD4+ T lymphocytes could result in an abnormal polyclonal activation of B lymphocytes with the production of auto-antibodies or the proliferation of previously dormant clones of autoreactive lymphocytes. It is possible that drugs could interfere with antigen presentation and the resulting polyclonal activation. However, no data are available to identify the involved mechanism at the molecular level: Could causative drugs alter MHC class II antigens to induce a direct activation of CD4+ T lymphocytes? Does the change affect the activation process of T or B lymphocytes? These pivotal questions remain unanswered.

Three examples of possible chemically-induced pseudo-GvH diseases

The Spanish toxic oil syndrome In May 1981, an epidemic of pulmonary diseases was seen in Madrid and the north-eastern provinces of Spain (Gómez de la Cámara *et al.*, 1997). The epidemic lasted until June 1984 and over 20 000 people were affected with over 1500 deaths. The epidemiological inquiry showed that an adulterated industrial oil illegally sold as oil for food was involved. Clinically, the disease presented with an acute phase associated with fever, cutaneous eruptions, interstitial pneumonitis, pleuropericarditis, gastrointestinal pain and eosinophilia, which developed approximately 2–3 weeks after oil ingestion. A few patients died at this stage from acute respiratory failure. Two months later, the chronic phase started with myalgia, arthralgia, thrombocytopenia, then Raynaud's phenomenon, Sjögren's syndrome and pulmonary hypertension, mimicking scleroderma. Patients died from neuromuscular disorders and malnutrition. Various immunological disorders were inconsistently reported: hypereosinophilia in the acute phase, decreased CD8+ T lymphocyte numbers in 75 per cent of patients in the chronic phase, auto-antibodies (essentially antinuclear antibodies), but no circulating immune complexes. The causative agent has never been identified. However, the role of microbial pathogens, heavy metals, mycotoxins, or pesticides could be ruled out. Based on the clinical pattern and observed immunological changes, the syndrome was suggested to be a pseudo-GvH disease, which could be due to the cyclisation derivative of aniline, 5-vinyl-2-oxazolidine-thione, formed during the manufacture of the adulterated oil.

Guillain–Barré syndrome due to zimeldine A second example is zimeldine, an antidepressant marketed in Sweden and several European countries in 1982 (Nilsson, 1983). In early clinical trials, approximately 2 per cent of patients developed fever, arthralgias, myalgias, cutaneous eruptions and digestive disorders. Later, peripheral neuropathies suggestive of the Guillain–Barré syndrome were described. The drug was withdrawn on 19 September 1983. Preclinical toxicity studies, in particular guinea-pig contact sensitisation assays, were unable to detect these adverse effects. However, later studies using the popliteal lymph node assay confirmed that a pseudo-GvH reaction was a likely mechanism, whereas conventional immunogenicity tests were all negative.

Eosinophilia–myalgia due to L-tryptophan In 1989, the first case reports of patients with a rare clinical syndrome, the 'eosinophilia–myalgia syndrome' were recorded in the USA. A survey conducted by the Center for Diseases Control in Atlanta quickly identified a causal relationship between the intake of products containing L-tryptophan and the onset of the syndrome. When the epidemic stopped in 1990, over 1500 patients with the syndrome had been detected in the USA, of whom 27 died. Patients with the syndrome were also observed in Canada, Japan, Australia and most European countries, although far fewer patients were affected. For instance, only 24 patients with this syndrome were officially recorded in France.

Three clinical stages were described (Belongia *et al.*, 1992): initially, patients presented with asthenia, fever, myalgias, and hypereosinophilia. Within two months, various complications developed, including peripheral neuropathy, myopathy, skin eruptions, and gastrointestinal disorders, in association with initial clinical symptoms. After six months, patients presented with a clinical condition mimicking scleroderma, with fibrosis and fibrotic lesions present in peripheral nerves, muscle fibres (e.g. in the heart), and vessels (e.g. lung vessels and coronary arteries). Interestingly, a number of immunological changes, with auto-antibodies as the commonest findings, were evidenced suggesting an immunological origin. No treatment proved effective, except corticosteroids to some extent.

Despite the widespread use of L-tryptophan, no safety problems had been suspected in relation to L-tryptophan intake prior to this epidemic. Interestingly, the Japanese firm Showa-Denko, a major world producer of L-tryptophan, changed the manufacturing process shortly before the beginning of the epidemic, and intake of L-tryptophan produced by this firm was documented or suspected in most, if not, all patients with the syndrome. Various contaminants or impurities were identified, such as the so-called peak 'E', later characterised as 1,1-ethylene,bis-tryptophan. However, the causative role of one particular contaminant could not be conclusively established.

The eosinophilia–myalgia syndrome associated with L-tryptophan had many similarities with the toxic oil syndrome: a well-characterised time course, the suspected involvement of contaminants previously not found, a clinical pattern evolving in three stages ending in a pathological condition similar to a systemic autoimmune disease, and various, if not inconsistent, immunological changes. Although uncertainties remain regarding the pathophysiology of both epidemics and are unlikely to be solved after so many years, the toxic oil syndrome and the eosinophilia–myalgia syndrome are clearly suggestive that autoimmune reactions induced by chemicals may affect a large number of human beings and cause significant morbidity and mortality.

Isolated auto-antibodies

The detection of auto-antibodies in asymptomatic patients is not uncommon (Tomer and Shoenfeld, 1988). The significance of this finding both from a pathogenetic point of view and in relation to drug treatment or toxic exposure is unknown. For instance, auto-antibodies have been identified in patients treated with a variety of cardiovascular drugs (Wilson, 1980) or in individuals exposed to silica (Steenland and Goldsmith, 1995). Auto-antibodies can also be detected in the sera of apparently healthy people. Auto-antibodies are also more frequent in association with pathological conditions, such as hypertension.

The physiological or pathological role of auto-antibodies is unknown, so that it is very difficult to propose a realistic management of asymptomatic patients with auto-antibodies. If the patient is a woman or is elderly, it is not certain whether cessation of drug treatment or chemical exposure should be recommended provided that the drug or chemical under consideration has not been reported to induce autoimmune reactions. Is it appropriate to stop drug therapy in patients who develop auto-antibodies when the drug is known to induce auto-antibodies and, as a rule, only very exceptionally patent autoimmune reactions?

References

ADAMS, L.E. and HESS, E.V. (1991) Drug-related lupus. *Drug Safety*, **6**, 431–449.

AMOS, H.E., LAKE, B.G. and ARTIS, J. (1978) Possible role of antibody specific for a practolol metabolite in the pathogenesis of the oculomucocutaneous syndrome. *Br. Med. J.*, **1**, 402–404.

ANHALT, G.J. (1989) Drug-induced pemphigus. *Semin. Dermatol.*, **8**, 166–172.

BEAUNE, P., DANSETTE, P.M., MANSUY, D., KIFFEL, L., FINCK, M., AMAR, C., *et al.* (1987) Human anti-endoplasmic reticulum autoantibodies appearing in a drug-induced hepatitis are directed against a human liver cytochrome P-450 that hydroxylates the drug. *Proc. Natl Acad. Sci. USA*, **84**, 551–555.

BEAUNE, P., PEYSSAYRE, D., DANSETTE, P., MANSUY, D. and MANNS, M. (1994) Auto-antibodies against cytochromes P450: role in human diseases. *Adv. Pharmacol.*, **30**, 199–245.

BELONGIA, E.A., MAYENO, A.N. and OSTERHOLM, M.T. (1992) The eosinophilia–myalgia syndrome and tryptophan. *Annu. Rev. Med.*, **12**, 235–256.

BIGAZZI, P.E. (1994) Autoimmunity caused by xenobiotics. *Toxicology*, **119**, 1–21.

BOURDI, M., LARREY, D., NATAF, J., BERNUAU, J., PESSAYRE, D., IWASAKI, M., *et al.* (1990) Anti-liver endoplasmic reticulum autoantibodies are directed against human cytochome P-450IA2. A specific marker of dihydralazine-induced hepatitis. *J. Clin. Invest.*, **85**, 1967–1973.

BOURGEOIS, P. and AESCHLIMANN, A. (1991) Drug-induced scleroderma. *Clin. Rheumatol.*, **5**, 13–20.

FERRARA, J.L.M. and DEGG, H.J. (1991) Graft-versus-host disease. *N. Engl. J. Med.*, **324**, 667–674.

GLEICHMANN, E., PALS, S.T., ROLINK, A.G., RADASZKIEWICZ, T. and GLEICHMANN, H. (1984) Graft-versus-host reactions: clues to the etiopathology of a spectrum of immunological diseases. *Immunol. Today*, **5**, 324–332.

GÓMEZ DE LA CÁMARA, A., ABAITUA BORDA, I. and POSADA DE LA PAZ, M. (1997) Toxicologists versus toxicological disasters: toxic oil syndrome, clinical aspects. *Arch. Toxicol.*, **suppl. 19**, 31–40.

GOOTLIEB, A.J. and WURZEL, H.A. (1974) Protein-quinone interaction: in vitro induction of indirect antiglobulin reactions with methyldopa. *Blood*, **43**, 85–97.

GRIEM, P., WULFERINCK, M., SACHS, B., GONZÁLEZ, J.B. and GLEICHMANN, E. (1998) Allergic and autoimmune reactions to xenobiotics: how do they arise? *Immunol. Today*, **19**, 133–141.

HALLA, J.T., FALLAHI, S. and KOOPMAN, W.J. (1984) Penicillamine-induced myositis. Observations and unique features in two patients and review of the literature. *Am. J. Med.*, **77**, 719–722.

HOLLAND, K. and SPIVAK, J.L. (1992) Drug-induced immunological disorders of the blood. In: *Clinical Immunotoxicology* (Newcombe, D.S., Rose, N.R. and Bloom, J.C., eds), pp. 141–153. New York: Raven Press.

KAMMÜLLER, M.E., BLOKSMA, N. and SEINEN, W. (1989) *Autoimmunity and Toxicology.* Amsterdam: Elsevier.

KARDESTUNCER, T. and FRUMKIN, H. (1997) Systemic lupus erythematosus in relation to environmental pollution: an investigation in an African-American community in North Georgia. *Arch. Environ. Health.*, **52**, 85–90.

KILBURN, K.H. and WARSHAW, R.H. (1994) Chemical induced autoimmunity. In: *Immunotoxicology and Immunopharmacology*, 2nd edition (Dean, J.H., Luster, M.I., Munson, A.E. and Kimber, I., eds), pp. 523–538. New York: Raven Press.

KIRTLAND, H.H., MOHLER, D.N. and HORWITZ, D.A. (1980) Methyldopa inhibition of suppressor-lymphocyte function. A proposed cause of autoimmune hemolytic anemia. *N. Engl. J. Med.*, **302**, 825–832.

MASSEY, J.M. (1997) Acquired myasthenia gravis. *Neurol. Clin.* **15**, 577–595.

NILSSON, B.S. (1983) Adverse reactions in connection with zimeldine treatment: a review. *Acta Psychiatr. Scand.*, **68**, suppl. 308, 115–119.

PERRY, R.M., CHAPLIN, H., CARMODY, S., HAYNES, C. and FREI, C. (1971) Immunologic findings in patients receiving methyldopa: a prospective study. *J. Lab. Clin. Med.*, **78**, 905–917.

PETER, J.B. and SHOENFELD, Y. (1996) *Autoantibodies*. Amsterdam: Elsevier.

PFEIFFER, R.W. and IRONS, R.D. (1985) Mechanisms of sulfhydryl-dependent immunotoxicity. In: *Immunotoxicology and Immunopharmacology*, 1st edition (Dean, J.H., Luster, M.I., Munson, A.E. and Amos, H., eds) pp. 255–262. New York: Raven Press.

PLADEVALL-VILA, M., DELCLOS, G.L., VARAS, C., GUYER, H., BRUGUES-TARRADELLAS, J. and ANGLADA-ARISA, A. (1996) Controversy of oral contraceptives and risk of rheumatoid arthritis: meta-analysis of conflicting studies and review of conflicting meta-analysis with special emphasis on analysis of heterogeneity. *Am. J. Epidemiol.*, **144**, 1–14.

PRICE, E.J. and VENABLES, P.J. (1995) Drug-induced lupus. *Drug Safety*, **12**, 283–290.

STEENLAND, K. and GOLDSMITH, D.F. (1995) Silica exposure and autoimmune diseases. *Am. J. Indust. Med.*, **28**, 603–608.

TOMER, Y. and SHOENFELD, Y. (1988) The significance of natural autoantibodies. *Immunol. Invest.*, **17**, 389–424.

UETRECHT, P. (1994) Current trends in drug-induced autoimmunity. *Toxicology*, **119**, 37–43.

VIAL, T., NICOLAS, B. and DESCOTES, J. (1994) Drug-induced autoimmunity: experience of the French Pharmacovigilance system. *Toxicology*, **119**, 23–27.

WILSON, J.D. (1980) Antinuclear antibodies and cardiovascular drugs. *Drugs*, **19**, 292–305.

WITTBRODT, E. (1997) Drugs and myasthenia gravis. An update. *Arch. Intern. Med.*, **157**, 399–408.

WONG, O. (1996) A critical assessment of the relationship between silicone breast implants and connective tissue diseases. *Regul. Toxicol. Pharmacol.*, **23**, 74–85.

YAMAUCHI, Y., LITWIN, A., ADAMS, L.E., ZIMMER, H. and HESS, E.V. (1975) Induction of antibodies to nuclear antigens in rabbits by immunization with hydralazine-human serum albumin conjugates. *J. Clin. Invest.*, **56**, 958–969.

7

Major Immunotoxicants

Many medicinal products, and industrial as well as environmental chemicals have been reported to induce immune changes in laboratory animals. However, a majority of these changes were noted in conditions which were not adequately controlled and/or selected on realistic grounds as regards the route, the duration and the magnitude of exposure, so that caution should be exercised when interpreting data obtained in early non-clinical immunotoxicity studies. Nevertheless, these findings indicate that the immune system is a target organ for toxic injury so that a systematic and thorough assessment of the immunotoxic potential of xenobiotics is certainly warranted in contrast to what is suggested by the current status of immunotoxicity regulations worldwide. In addition to histological and functional immune changes, hypersensitivity reactions and to a much lesser extent autoimmune reactions have been described. Except for hypersensitivity reactions, data in humans are scarce and efforts should be paid to investigating better the immunotoxic effects of xenobiotics in man.

The scope of this chapter is limited to a brief overview of major known or suspected immunotoxicants. Comprehensive reviews are available elsewhere (for instance: Dayan et al., 1990; Dean et al., 1994; Descotes, 1999; Mitchell et al., 1990; Sullivan, 1989).

Medicinal products

Many medicinal products have been suggested or shown to induce functional immune changes, hypersensitivity reactions, and/or autoimmune reactions (Descotes, 1990). However, drug-induced immune-mediated clinical adverse effects, with the exception of hypersensitivity reactions, have been uncommonly recorded. Possible explanations to account for these apparent discrepancies are that histological and functional immune changes are not necessarily indicative of overt immunotoxicity (as electrocardiographic changes are not necessarily indicative of cardiotoxicity); that few reliable assays are available for the diagnosis of drug-induced immune-mediated adverse effects in humans; and that under-reporting is likely to be very common due to the lack of awareness that the immune system can indeed be a target organ of toxicity.

Antimicrobials

A number of antibiotics and antimicrobials have been shown to induce immune changes involving both non-specific host defences and specific immune responses (Hauser and Remington, 1982; Jeljsazewicz and Pulverer, 1986), even though the relevance of these changes remains ill-understood today, as no significant differences in the therapeutic efficacy or the development of adverse reactions have been clearly demonstrated to be related to the immunotoxic potential of antimicrobials. Although class-related immunotoxic properties have been identified, the individual chemical structure is the key determinant (Labro, 1996). Another critical aspect is the limited amount of data for elucidating the mechanism of interferences between antimicrobials and the immune system or host defences.

Hypersensitivity reactions are relatively common and potentially severe, even life threatening, with several major groups of antimicrobials. In contrast, autoimmune reactions are very rare.

Direct immunotoxicity

A large number of experimental studies in the 1980s were devoted to investigating the interference of antimicrobials with the immune response towards microbial pathogens. From a theoretical point of view, it is logical to propose that antimicrobials should not impair the host's defence mechanisms against pathogens, particularly in immuno-compromised patients. The relevance of findings obtained *in vitro* or in healthy animals or humans is unclear, and the lack of overt therapeutic benefit associated with 'immuno-enhancing' antimicrobials or adverse consequences associated with 'immunodepressive' antimicrobials resulted in the nearly total cessation of such investigations in recent years.

Tetracyclines, aminoglycosides, chloramphenicol, and several antifungal drugs, such as amphotericin B, ketoconazole and niridazole, have been shown to impair markedly the phagocytic and chemotactic functions of neutrophils. In contrast, penicillins, cephalosporins and macrolides have limited, or no effects. As regards specific immune responses, chloramphenicol, which was found to exert marked immunosuppressive effects *in vivo* and was even proposed as an adjunct to immunosuppressive therapy, rifampicin, tetracyclines, macrolides and a few cephalosporin derivatives have all been reported to exert negative effects.

A recent survey of papers published between 1987 and 1994 concluded that the amount of data is often too small to draw conclusions on the immunomodulating or immunotoxic properties of most antimicrobials (Van Vlem *et al.*, 1996). Only three antimicrobials with immuno-enhancing properties were identified, namely cefodizime, clindamycin and imipenem, and eight with immunodepressive properties (erythromycin, roxithromycin, cefotaxime, tetracycline, rifampicin, gentamicin, teicoplanin and ampicillin).

Hypersensitivity reactions

Antimicrobials are the most frequent causes of hypersensitivity reactions induced by medicinal products, even though it can be difficult to ascertain whether a patient is really allergic (Boguniewicz, 1995).

Beta-lactam antibiotics are the most frequent cause of immediate drug-induced immune allergic reactions. The estimated prevalence of penicillin allergy is 1–2 per cent. Acute IgE-mediated reactions include anaphylaxis, angioedema and urticaria, but other adverse reactions, such as immune allergic haemolytic anaemia, serum sickness-like reactions and contact dermatitis, are also described (DeShazo and Kemp, 1997). The penicilloyl hapten is the major determinant, which accounts for approximately 95 per cent of clinical reactions. The probability that patients with a history suggestive of penicillin allergy will develop an adverse clinical reaction upon readministration, can be determined by skin testing. Patients with a positive skin test to minor determinants are considered to have a far greater risk of developing anaphylaxis. Penicillin-induced anaphylaxis has not been reported in patients with negative skin tests, whereas 1 to 10 per cent of patients with no history of a clinical reaction, but positive skin tests, develop anaphylaxis. Oral and intravenous desensitisation protocols have been used successfully. Oral protocols are less likely to induce anaphylaxis, but pruritus and skin rash develop in 5 per cent of patients. The mechanism of desensitisation is not fully elucidated. The risk for anaphylaxis upon readministration of penicillin decreases with time in patients with a history of allergic reaction.

All other beta-lactam antibiotics cause immuno-allergic reactions, but less frequently than penicillin. Immunological cross-reactivity is high between penicillins and carbapenems, less to cephalosporins (and particularly the so-called third generation cephalosporins) and least to monobactams. Because the haptenic determinants are unknown, the results of skin tests should be used with caution.

Immune-allergic reactions are also common in patients treated with sulphonamides (DeShazo and Kemp, 1997). A generalised maculopapuplar rash is the most frequent clinical reaction, but severe mucocutaneous reactions, such as the Stevens–Johnson syndrome, can occur. As sulphonamides are metabolised by N-acetylation, slow acetylators are genetically more prone to generate oxygenated metabolites, which are normally neutralised by glutathione reductase. Decreased activity of this enzyme in HIV-infected individuals could account for the much greater prevalence of immuno-allergic reactions to sulphonamides, as well as to other drugs metabolised by N-oxidation, such as dapsone and rifampin.

In contrast, tetracyclines, macrolides and quinolones seldom cause immuno-allergic hypersensitivity reactions.

Autoimmune reactions

Only lupus syndromes have been reported, but quite exceptionally, in association with a few antimicrobials, such as tetracyclines or quinolones. However, a number of lupus syndromes have surprisingly been reported recently with the tetracycline derivative, minocycline, although it has been in use for over 30 years (Knowles *et al.*, 1996). Nitrofurantoin, sulphasalazine and the antituberculosis drug isoniazid are among the most commonly incriminated antimicrobials causing lupus syndromes.

Antiepileptic drugs

A number of immune changes and immune-mediated adverse reactions have been reported in patients treated with a variety of antiepileptic drugs (Cereghino, 1983; De Ponti *et al.*, 1993).

IgA deficiency used to be considered the most frequent immunological disorder associated with antiepileptic treatment. However, IgA deficiency (namely serum IgA levels below 0.5 mg/ml) is relatively common in the general population (1 in 500–700 individuals), so that it is always difficult to establish a causal relationship between IgA deficiency and antiepileptic therapy, particularly as no pretreatment IgA level measurement is usually performed. Anyway, most major antiepileptics (carbamazepine, valproic acid and especially diphenylhydantoin) were reportedly associated with IgA deficiency. The lack of data on IgA levels in patients treated with the most recent antiepileptics might in fact suggest that IgA deficiency associated with antiepileptics was unduly overemphasised. Several antiepileptic drugs, such as carbamazepine, diphenylhydantoin and phenobarbital, in contrast to sodium valproate, have been shown to exert immunodepressive properties. However, the clinical consequences of these findings remain to be established.

Epileptic patients treated with diphenylhydantoin or carbamazepine, sometimes develop benign lymphadenopathy with adenopathies in the cervical area, hyperpyrexia, cutaneous eruptions, arthritis, pneumonitis, hepatomegaly or hepatitis, so the diagnosis of hypersensitivity reaction or malignant haemopathy due to treatment was proposed. However, the evolution to malignant lymphoma was rarely described. The term 'drug hypersensivity syndrome' was coined to describe these reactions, in the pathophysiology of which IL-5 was suggested to play a central role. The recent antiepileptic lamotrigine was reported to induce frequent skin reactions, and sometimes severe toxidermias, such as the Stevens–Johnson syndrome, or toxic epidermal necrolysis.

Finally, it was recognised years ago that antiepileptic therapy can be associated with lupus syndromes. Many reports involved diphenylhydantoin, carbamazepine and the older derivative trimethadione. Lupus syndrome does not seem to occur in patients treated with the most recent derivatives.

Anti-inflammatory drugs and antirheumatics

Non-steroidal anti-inflammatory drugs (NSAIDs), in particular aspirin, have been the subject of many investigations in the past 30 years to determine whether they can modulate immune functions (Goodwin, 1985). Despite the bulk of published results, it is difficult to draw firm conclusions, except that a marked influence on immune responsiveness is unlikely.

NSAIDs have varied chemical structures, and several, but not all derivatives induce immuno-allergic hypersensitivity reactions (Hoigné and Szczeklik, 1992). Phenylbutazone and oxyphen(yl)butazone have been reported to cause a variety of potentially severe immune-mediated adverse reactions, such as agranulocytosis and toxic epidermal necrolysis. A major adverse effect of NSAIDs is the intolerance syndrome, described in Chapter 5 of this volume. In fact, intolerance to NSAIDs is a prototypic pseudo-allergic reaction and affected patients can develop similar and recurring clinical symptoms when administered one of several structurally unrelated NSAIDs. Minor analgesics have limited potential for inducing immunotoxicity. Glafenin was withdrawn from the market because of anaphylatic shock, which actually had been documented in few patients as compared to mild to moderate and spontaneously recovering 'pseudo-allergic' shocks described many years ago. Paracetamol is very unlikely to exert immunotoxic adverse effects.

In contrast to NSAIDs, corticosteroids have long been reported to be potent immunotoxicants (Kass and Finland, 1953; Thomas, 1952). Adverse effects related to immunosuppression associated with corticosteroid therapy were indeed described shortly after the introduction of corticosteroids into the clinical setting. Infectious complications are common and directly related to the dose level and duration of treatment exposure. Opportunistic or particularly severe infections, such as malignant varicella, have been reported. In contrast, lymphomas seem to be extremely rare and no other neoplasias have been suspected to be associated with corticosteroid therapy, presumably because of the many severe and potentially treatment-limiting non-immune-mediated adverse effects associated with corticosteroid therapy. Corticosteroids have recently been recognised as a cause of hypersensitivity reactions, for instance contact dermatitis and anaphylactic shock (Dooms-Goosens *et al.*, 1989). Recent debate on the sensitising potency of corticosteroids was due to the finding that drug additives included in the formulation of several corticosteroids, such as sulphites, could not account for all reported adverse reactions.

Antirheumatics seem to be relatively devoid of direct immunotoxic effects, although most exert immunopharmacological properties which could presumably account for their therapeutic efficacy (Bálint and Gergely, 1996). Hypersensitivity reactions, including cutaneous eruptions, blood disorders, and eosinophilic pneumonitis, have been described with gold salts, whereas penicillamine and to a lesser extent, tiopronine and pyritinol, can cause autoimmune reactions, such as myasthenia, pemphigus, polymyositis, and the lupus syndrome.

Cardiovascular drugs

Very limited information is available regarding the direct immunotoxicity of cardiovascular drugs. β-Adrenergic inhibitors (β-blockers) have been inconsistently shown to decrease immune responses, whereas the angiotensin-converting enzyme inhibitor captopril was suggested to have immunostimulating properties. The calcium antagonists verapamil, nifedipine and dilthiazem have been shown to exert immunodepressive effects of limited significance in treated patients.

A variety of immune-mediated clinically significant adverse effects have been described in patients treated with antihypertensive drugs. Inhibitors of angiotensin-converting enzyme produce angioedema in 0.1–0.2 per cent of patients (Vleeming *et al.*, 1998). Patients on β-blockers have been shown to be at greater risk of developing severe anaphylactic reactions (Lang, 1995).

Autoimmune reactions are the most common immunotoxic adverse effects. An enormous number of lupus syndromes have been described in association with hydralazine. Nearly every β-blocker on the market has been reported to induce lupus, with acebutolol and practolol as the most commonly incriminated derivatives. Other autoimmune reactions include pemphigus associated with captopril, autoimmune haemolytic anaemia associated with α-methyldopa, and isolated reports of lupus syndrome associated with diuretics.

Hormones

Hormones and hormone derivatives have consistently been shown to influence immune responsiveness, either positively or negatively depending on the hormone under scrutiny, but also the initial hormonal status. Overall, oestrogens, progesterone and androgens have immunodepressive activities (Michael and Chapman, 1990).

A causal relationship between the use of oral contraceptives and the development of rheumatoid arthritis or systemic lupus erythematosus has been suggested and debated. The most recent data indicate that the risk, if present, is probably very slight (Pladevall-Vila *et al.*, 1996). Thromboembolic accidents in at least some contraceptive users were reported to be associated with the presence of specific anti-oestrogen antibodies in their sera (Beaumont *et al.*, 1982).

Psychotropic drugs

Many reports focused on the possible immunotoxicity of psychotropic drugs and despite the bulk of information, it is still difficult to draw firm conclusions as to whether and which psychotropic drugs are actually immunotoxic. Experimental data obtained either *in vitro* or *in vivo* in laboratory animals, suggested that most psychotropic drugs had the capacity to induce changes in immune responses. However, such changes could seldom be linked to clinical immunotoxic consequences. Hypersensitivity reactions are relatively uncommon, as are autoimmune reactions.

Among major tranquilisers, the phenothiazine derivatives, including chlorpromazine and promethazine, have been shown to exert marked immunosuppressive effects. Promethazine was even once considered as a possible adjunct to immunosuppressive therapy in transplant patients (Orlowski *et al.*, 1983). Chlorpromazine was also recorded as the likely cause of lupus syndromes (Pavlidakey *et al.*, 1985). Auto-antibodies in the sera of asymptomatic patients under phenothiazine treatment are not uncommon, but the clinical relevance of this finding is unknown. Other major tranquilisers, such as haloperidol and risperidone, have been shown to modulate immune responses, but no clinical immunotoxic consequences have so far been identified in treated patients.

In contrast to phenothiazine derivatives, tricyclic antidepressants and benzodiazepines are generally considered devoid of immunotoxicity, even though binding of tricyclic antidepressants to surface membrane sites of lymphocytes has been identified as well as binding of benzodiazepines on peripheral receptors. In general, experimental studies could only evidence some depression of immune responsiveness with high dose levels of these drugs. Several studies showed that prenatal exposure to the benzodiazepine diazepam was associated with impaired immune responsiveness in adult mice, but the clinical relevance of these findings is unknown (Schlumpf *et al.*, 1992).

Nomifensine was shown to induce autoimmune haemolytic anaemia which led to its withdrawal from the market, and iproniazid to induce immunotoxic hepatitis. Conflicting effects of lithium salts on specific immune responses have been reported, but a stimulation of leukocyte functions was consistently observed, which could account for the reported exacerbation of psoriasis in several lithium-treated patients.

Vaccines

Adverse reactions to vaccines are not so uncommon as generally held. A majority of these adverse reactions involve the immune system, and anaphylaxis has repeatedly been reported to be a typical, although fairly uncommon complication of most vaccines (Kobayashi, 1995). Mild reactions include local inflammatory reactions at the site of injection, fever, and skin eruptions. More severe, but fortunately rare, reactions include anaphylactic shock, lymphadenitis, and vasculitis. The causative role of preservatives, such as thimerosal, adjuvants, such as aluminium hydroxide, and impurities, such as egg proteins or antibiotics, should not be overlooked. The role of vaccines in the development of autoimmune reactions is a matter of growing debate.

Medical devices

Medical devices, such as implants, drug delivery systems, and extracorporeal devices, include any item promoted for a medical purpose that does not rely on chemical action to achieve its intended effect. The possible immunotoxicity of medical devices has rarely been investigated (Rodgers *et al.*, 1997). However, hypersensitivity reactions to the titanium hip prosthesis resulting in further failure of the prosthesis are a good example of possible immunotoxic adverse effects associated with medical devices.

Use of the silicon prosthesis for breast replacement or breast enlargement, or silicon gel to delete wrinkles was reportedly and conflictingly associated with more frequent systemic autoimmune reactions, especially scleroderma-like diseases (Wong, 1996). Although no causal link between implantation of medical devices and the onset of disease could be firmly established, such case reports suggest that medical devices may alter the immune responsiveness of implanted patients and induce immune-mediated adverse effects.

Industrial and environmental pollutants

Data related to the immunotoxicity of industrial and environmental chemicals have been essentially obtained in laboratory animals. Although the immunotoxicity potential of a number of compounds has been investigated, interest has focused on several compounds, such as dioxin and polycyclic aromatic hydrocarbons, metals such as mercury, lead and cadmium, and a few pesticides. In addition, the immunotoxicity of industrial and environmental chemicals was largely, if not exclusively, considered from the perspective of immunosuppression, so that few data are available regarding their potential for inducing hypersensitivity and autoimmune reactions.

Heavy metals

Many experimental works have dealt with the influence of heavy metals on the immune system, but have often produced conflicting results (Dayan *et al.*, 1990; Zelikoff and Thomas, 1998). Therefore, even today, it is still very difficult to draw a realistic picture of how immunotoxic heavy metals are, either at the workplace or in the environment. In addition, low exposures to metals, such as cadmium and lead, have sometimes been shown to exert immunostimulating effects in contrast to higher exposures, but these opposed effects have not been carefully assessed from the perspective of risk assessment.

Immunosuppression

The majority of heavy metals, especially lead, mercury, cadmium, and organotins, were found to be immunodepressive in laboratory animals. Tributyltin oxide is a prototypic immunotoxicant (Verdier *et al.*, 1991). Rats fed tributyl oxide were found to develop thymic atrophy at dose levels which did not induce any other toxic injury or functional immune change. Tributyltin oxide was the first environmental chemical for which maximal accepted concentrations in the environment were estimated based on immunotoxicity data. However, it is not always possible to show a relation between metal-induced alterations in humoral and/or cell-mediated immunity, and the impaired resistance towards experimental infections, even though interlaboratory validation studies, such as the US National Toxicology Program study, obtained such relations with a panel of pre-selected compounds. Selenium is the only metal with immunostimulating properties, although increasing the dose quickly results in immunodepression (Kiremidjian-Schumacher and Stotzky, 1987).

Several factors account for the conflicting results of published studies. The importance of exposure levels has already been stressed: low-level exposures may enhance, whereas higher exposures decrease immune responses and intermediate exposures possibly exert no effect at all, as shown with arsenic, cadmium, lead, or selenium. The time schedule of exposure with respect to antigen administration or experimental infection is another critical factor: low-level exposures to lead prior to injection of microbial pathogens enhance murine host resistance, whereas high-level exposures when microbial pathogens are injected, depress host resistance (Laschi-Loquerie *et al.*, 1987). The chemical species is another critical factor: past studies focused on metals as one single chemical species, although early data showed that different salts of the same metal, e.g. lead, can exert opposed effects on the same immune parameter (Descotes *et al.*, 1984).

As with most industrial and environmental chemicals, very limited information is available in humans, so that it is largely unknown whether animal findings following exposure to metals and metal salts, can be extrapolated to human beings. In addition, when immune function changes have been identified, such as depressed neutrophil functions in workers chronically exposed to lead (Bergeret *et al.*, 1990), the clinical relevance of such changes was not investigated.

Autoimmunity

Autoimmunity induced by heavy metals has rarely been reported, except in particular, if not to say, artificial experimental conditions. A polyclonal activation of B lymphocytes by lead or mercury has sometimes been described. The experimental autoimmune glomerulonephritis to mercuric chloride in Brown-Norway rats has no human counterpart, despite the largely undocumented claim that mercuric salts could induce autoimmunity in humans. Field studies in workers consistently confirmed the lack of adverse influence of mercury exposure on the immune system, and particularly the lack of auto-antibodies (Langworth *et al.*, 1992). The issue of whether mercury dental amalgams can exert immunotoxic effects is a matter of controversy.

Hypersensitivity

Every metal, essentially when inhaled, can cause pulmonary hypersensitivity reactions. A direct activation of macrophages with the subsequent release of IL-1 and TNF-α could account for these reactions, of which zinc welder fever is a typical example (Blanc *et al.*, 1991).

Beryllium deserves special attention, as it can induce chronic pulmonary granulomatosis or berylliosis, which irreversibly evolves to respiratory failure. The clinical manifestations of berylliosis, which are similar to those of sarcoidosis, are probably related to an immune-mediated mechanism. The lymphocyte proliferation test is generally considered to be a reliable tool to follow up beryllium-exposed workers and detect sensitisation to beryllium at an early stage (before the onset of berylliosis), even though this position has sometimes been debated.

Platinum, chromium and several other metal salts can induce anaphylactic reactions. Other metals, such as nickel or chromium, are frequent causes of allergic contact dermatitis.

Hydrocarbons

Aliphatic hydrocarbons, such as trichloroethane and trichloroethylene, have limited, if no influence on the immune system, whereas most aromatic hydrocarbons are potent immunotoxicants in laboratory animals.

Aromatic hydrocarbons

Benzene is leukaemogenic in man, but the mechanism is only partly understood. Inhalation of very small benzene concentrations was found to be immunodepressive in rodents. The parent molecule is unlikely to be directly toxic, and metabolites suspected to play a key role in the toxicity of benzene, namely parabenzoquinone and hydroquinone, have been shown to be markedly immunosuppressive (Smialowicz, 1996). Available studies suggest that toluene and styrene are unlikely to have marked effects on immunological function in animals and man.

Polycyclic aromatic hydrocarbons

Polycyclic aromatic hydrocarbons (PAHs), which are potent carcinogens and mutagens, are also markedly depressive of the humoral and/or cell-mediated immune responses in rodents and humans *in vivo* and/or *in vitro*, resulting in increased susceptibility of the host towards microbial infections and implanted tumours (Ladics and White, 1996). A majority of PAHs were shown to decrease NK cell activity in rodents. Benzo[a]pyrene, methylcholanthrene, and 7, 12-dimethyl-benz[a]thracene have been the most extensively investigated PAHs.

Halogenated organic compounds

Halogenated organic compounds consist of a variety of chemical families, such as polychlorinated dibenzo-*p*-dioxin, polychlorinated dibenzofurans, polychlorinated and polybrominated biphenyls, and organochlorine insecticides. Although the acute toxicity is usually relatively low in humans, they persist for a long time in nature due to their high lipophilicity and resistance to biological degradation, so that they accumulate within the food chain and the environment.

Biphenyls

In 1973, cattle feed was accidentally contaminated by brominated biphenyls in Michigan. Farmers who consumed dairy products from contaminated cattle developed more frequent immunological changes, such as depressed lymphocyte proliferative responses to mitogens, than matched controls from Wisconsin (Bekesi *et al.*, 1978). This accidental contamination was suggested to account for Hodgkin's disease in this population. Fifteen years later, solid tumours were suggested to be more frequent in the contaminated population than in the Wisconsin matched controls, but these preliminary findings were never confirmed or negated.

An accidental contamination of rice oil by polychlorinated biphenyls (PCBs) used as coolers, was observed in Japan in 1968 (Kuratsune *et al.*, 1996). More than 1800 Japanese were affected by the disease ('Yusho disease') with bronchitis and low immunoglobulin levels as the most common findings. In fact, contamination of rice oil was later identified to be due more to polychlorinated dibenzofurans than to PCBs. Another contamination of the food chain by dibenzofurans and other PCB-related impurities was reported in Taiwan in 1979 ('Yu-Cheng disease'). Patients developed acne-like skin disorders, liver disorders, a depression of delayed-type hypersensitivity response and reduced immunoglobulin levels up to one year after exposure.

In laboratory animals, biphenyls are markedly suppressive of the humoral immunity, more than the cell-mediated immunity. Susceptibility to experimental infections and resistance to implanted tumours were also shown to be decreased in exposed rodents.

Dioxin

The immunotoxic potential of dioxin has been the matter of an enormous number of experimental studies. Dioxin is one of the most potently immunosuppressive chemicals in rodents (Kerkvliet, 1994). It was shown to impair humoral and even more cell-mediated immunity, the susceptibility towards experimental infections and implanted tumours. By contrast, NK cell activity and macrophage functions are not markedly impaired. Dioxin can bind stereospecifically and irreversibly to an intracellular receptor, the Ah receptor, which is found on target cells, especially maturing thymocytes. Alterations in the maturation and differentiation of T lymphocytes resulting from dioxin binding to the Ah receptor is the widely held mechanism of dioxin immunotoxicity. Human data, although scarce, hardly confirmed animal findings: immune changes, such as reduced cutaneous response to recall antigens and impaired or increased lymphocyte proliferation, were sometimes reported, but no excess of clinical disease was found.

Pesticides

Although safety issues related to pesticide exposure are a matter of continuing concern, relatively limited data are available regarding their possible impact on the immune system (Repetto and Baliga, 1996; Thomas *et al.*, 1990), particularly in man (Vial *et al.*, 1996). Contact dermatitis has been described, but the incidence may actually be underestimated.

Chronic exposure to DDT, which caused depressed immune responses in rodents, did not change vaccinal responses in exposed children. The carbamate derivative aldicarb was a matter of vivid debate when it was reported that women exposed to small levels of aldicarb via the drinking water, had lower CD4[+] T lymphocytes. This 'chemical cause of AIDS', as the media exaggeratedly put it, was ruled out by further studies showing the lack of immunotoxic potential of aldicarb in mice. The systematic study of pesticides to evidence possible immunosuppressive effects in Wistar rats produced a majority of negative results (Vos *et al.*, 1983). However, due to their marked general toxicity and their widespread use, pesticides remain a matter of concern for immunotoxicologists.

Air pollutants

Many air pollutants, such as ozone, nitrogen oxides and sulphur hydrogen, exert adverse effects on the non-specific defence mechanisms of the host, particularly on alveolar macrophages. A direct immunotoxic effect on specific immune responses is possible, but remains to be clearly established (Selgrade and Gilmour, 1994). Nevertheless, resistance to experimental infections was consistently shown to be decreased following inhalation exposure to major air pollutants.

The role of air pollution in respiratory allergy and asthma is a matter of extensive investigation (Dybing *et al.*, 1996). Several comparative studies, for example, between former Eastern and Western Germanies, or between Sweden and Balkan countries, showed that decreased levels of air pollution are associated with more frequent asthma and respiratory allergies, whereas more pronounced levels of air pollution are associated with more frequent respiratory infections, suggesting a dose-related, but qualitatively different impact of air pollutants on the host's resistance.

Addictive drugs and chemicals

That illicit use of recreative or addictive drugs is associated with frequent and potentially severe infectious complications has long been recognised, and was initially thought to be largely due to unhealthy conditions of use/exposure. However, it is now well established that most addictive drugs and habits, for example, heroin, cocaine, marijuana, alcohol, and smoking, have the capacity to induce marked changes in immune function resulting in impaired antimicrobial resistance (Friedman *et al.*, 1991; Watson, 1990, 1993).

References

BÁLINT, G. and GERGELY, P. (1996) Clinical immunotoxicity of antirheumatic drugs. *Inflamm. Res.*, **45**, S91–S95.

BEAUMONT, V., LEMORT, N. and BEAUMONT, J.L. (1982) Oral contraception, circulating immune complexes, antiethinylestradiol antibodies and thrombosis. *Am. J. Reprod. Immunol.*, **2**, 8–12.

BEKESI, J.G., HOLLAND, J.F., ANDERSON, H.A., FISHBEIN, A.S., ROM, W., WOLF, M.S. and SELEKOFF, I.J. (1978) Lymphocyte function of Michigan dairy farmers exposed to polybrominated biphenyls. *Science*, **199**, 1207–1209.

BERGERET, A., POUGET, E., TEDONE, R., MEYGRET, T., CADOT, R. and DESCOTES, J. (1990) Neutrophil functions in lead-exposed workers. *Hum. Exp. Toxicol.*, **9**, 231–233.

BLANC, P., WONG, H., BERNSTEIN, M.S. and BOUSHEY, H.A. (1991) An experimental human model of metal fume fever. *Ann. Intern. Med.*, **114**, 930–936.

BOGUNIEWICZ, M. (1995) Adverse reactions to antibiotics. Is the patient really allergic? *Drug Safety*, **13**, 273–280.

CEREGHINO, J.J. (1983) Immunological aspects of epilepsy and antiepileptic drugs. In: *Chronic Toxicity of Antiepileptic Drugs* (Oxley J., ed.), pp. 251–259. New York: Raven Press.

DAYAN, A.D., HERTEL, R.F., HELSELTINE, E., KAZANTZIS, G., SMITH, E.M. and VAN DER VENNE, M.T. (1990) *Immunotoxicity of Metals and Immunotoxicology*. Proceedings of an International Workshop. New York: Plenum Press.

DEAN, J.H., LUSTER, M.I., MUNSON, A.E. and KIMBER, I. (1994) *Immunotoxicology and Immunopharmacology*, 2nd edition. New York: Raven Press.

DE PONTI, F., LECCHINI, S., COSENTINO, M., CASTELLETTI, C.M., MALESCI, A. and FRIGO, G.M. (1993) Immunological adverse effects of anticonvulsants. What is their clinical relevance? *Drug Safety*, **8**, 235–250.

DESCOTES, J. (1990) *Drug-Induced Immune Diseases*. Amsterdam: Elsevier.

DESCOTES, J. (1999) *Immunotoxicology of Drugs and Chemicals: Experimental and Clinical Perspectives*, 3nd edition. New York: Elsevier.

DESCOTES, J., LASCHI-LOQUERIE, A., TACHON, P. and EVREUX, J.C. (1984) Comparative effects of various lead salts on delayed hypersensitivity. *J. Appl. Toxicol.*, **4**, 265–266.

DeSHAZO, R.D. and KEMP, S.F. (1997) Allergic reactions to drugs and biologic agents. *JAMA*, **278**, 1895–1906.

DOOMS-GOOSENS, A., DEGREEF, H.J., MARIEN, K.J. and COOPMAN, S.A. (1989) Contact allergy to corticosteroids: a frequently missed diagnosis? *J. Am. Acad. Dermatol.*, **21**, 538–545.

DYBING, E., LØVIK, M. and SMITH, E. (1996) Environmental chemicals and respiratory hypersensitization. Special issue. *Toxicol. Letters*, **86**, 57–222.

FRIEDMAN, H., SPECTER, S. and KLEIN, T.W. (1991) *Drugs of Abuse, Immunity and Immunodeficiency*. New York: Plenum Press.

GOODWIN, J.S. (1985) Immunologic effects of nonsteroidal anti-inflammatory agents. *Med. Clin. N. Am.*, **69**, 793–804.

HAUSER, W.E. and REMINGTON, J.S. (1982) Effects of antibiotics on the immune response. *Am. J. Med.*, **72**, 711–716.

HOIGNÉ, R.V. and SZCZEKLIK, A. (1992) Allergic and pseudoallergic reactions associated with nonsteroidal anti-inflammatory drugs. In: *AINS – A Profile of Adverse Effects* (Borda, I.T. and Koff, R.S, eds), pp. 157–184. Philadelphia: Hanley & Belfus.

JELJASZEWICZ, J. and PULVERER, G. (1986) *Antimicrobial Agents and Immunity*. London: Academic Press.

KASS, E.H. and FINLAND, M. (1953) Adrenocortical hormones in infection and immunity. *Annu. Rev. Microbiol.*, **7**, 361–388.

KERKVLIET, N.I. (1994) Immunotoxicology of dioxins and related chemicals. In: *Dioxins and Health* (Schecter, A., ed.), pp. 199–225. New York: Plenum Press.

KIREMIDJIAN-SCHUMACHER, L. and STOTZKY, G. (1987) Selenium and immune responses. *Environ. Res.*, **42**, 277–303.

KNOWLES, S.R., SHAPIRO, L. and SHEAR, N.H. (1996) Serious adverse reactions induced by minocycline. Report of 13 patients and review of the literature. *Arch. Dermatol.*, **132**, 934–939.

KOBAYASHI, R.H. (1995) Vaccinations. *Immunol. Allergy Clin. N. Am.*, **15**, 553–565.

KURATSUNE, M., YOSHIMURA, H., HORI, H., OKUMURA, M. and MASUDA, Y. (1996) *Yusho. A Human Disaster caused by PCBs and Related Compounds*. Fukuoka: Kyushu University Press.

LABRO, M.T. (1996) Immunomodulatory actions of antibacterial agents. *Clin. Immunother.*, **6**, 454–464.

LADICS, G.S. and WHITE, K.L. (1996) Immunotoxicity of polyaromatic hydrocarbons. In: *Experimental Immunotoxicology* (Smialowicz, R.J. and Holsapple, M.P., eds), pp. 331–350. Boca Raton: CRC Press.

LANG, D.M. (1995) Anaphylactoid and anaphylactic reactions. Hazards of β-blockers. *Drug Safety*, **12**, 299–304.

LANGWORTH, S., ELINDER, C.G., SUNDQUIST, K.G. and VESTERBERG, O. (1992) Renal and immunological effects of occupational exposure to inorganic mercury. *Br. J. Indust. Med.*, **49**, 394–401.

LASCHI-LOQUERIE, A., EYRAUD, A., MORISSET, D., SANOU, A., TACHON, P., VEYSSEYRE, C. and DESCOTES, J. (1987) Influence of heavy metals on the resistance of mice toward infection. *Immunopharmacol. Immunotoxicol.*, **9**, 235–242.

MICHAEL, S.D. and CHAPMAN, J.C. (1990) The influence of the endocrine system on the immune system. *Immunol. Allergy Clin. N. Am.*, **10**, 215–233.

MITCHELL, J.A., GILLAM, E.M.J., STANLEY, L.A. and SIM, E. (1990) Immunotoxic side-effects of drug therapy. *Drug Safety*, **5**, 168–178.

ORLOWSKI, T., NIEBULOWICZ, J., GÓRSKI, A., GRADOWSKA, L., JUSKOWA, J., KLEPACKA, J., *et al.* (1983) A controlled prospective long term trial of promethazine (PM) as an adjuvant immunosuppressant in 102 cadaver graft recipients. *Transplant. Proc.*, **15**, 557–559.

PAVLIDAKEY, G.P., HASHIMOTO, K., HELLER, G.L. and DANESHAR, S. (1985) Chlorpromazine-induced lupus-like disease. Case report and review of the literature. *J. Am. Acad. Dermatol.*, **13**, 109–115.

PLADEVALL-VILA, M., DELCLOS, G.L., VARAS, C., GUYER, H., BRUGUES-TARRADELLAS, J. and ANGLADA-ARISA, A. (1996) Controversy of oral contraceptives and risk of rheumatoid arthritis: meta-analysis of conflicting studies and review of conflicting meta-analysis with special emphasis on analysis of heterogeneity. *Am. J. Epidemiol.*, **144**, 1–14.

REPETTO, R. and BALIGA, S.S. (1996) *Pesticides and the Immune System*. Washington DC: World Resources Institute.

RODGERS, K., KLYKKEN, P., JACOBS, J., FRONDOZA, C., TOMAZIC, V. and ZELIKOFF, J. (1997) Immunotoxicity of medical devices. *Fund. Appl. Toxicol.*, **36**, 1–14.

SCHLUMPF, M., PARMAR, R., SCHREIBER, A., RAMSEIER, H.R., BÜTIKOFER, E., ABRIEL, H., *et al.* (1992) Nervous and immune system as targets for developmental effects of benzodiazepines. A review of recent studies. *Dev. Pharmacol. Ther.*, **18**, 145–158.

SELGRADE, M.K. and GILMOUR, I. (1994) Effects of gaseous air pollutants on immune responses and susceptibility to infectious and allergic disease. In: *Immunotoxicology and Immunopharmacology*, 2nd edition (Dean, J.H., Luster, M.I., Munson, A.E. and Kimber, I., eds), pp. 395–412. New York: Raven Press.

SMIALOWICZ, R.J. (1996) Immunotoxicity of organic solvents. In: *Experimental Immunotoxicology* (Smialowicz, R.J. and Holsapple, M.P., eds), pp. 307–330. Boca Raton: CRC Press.

SULLIVAN, J.B. (1989) Immunological alterations and chemical exposure. *Clin. Toxicol.*, **27**, 311–343.

THOMAS, L. (1952) The effects of cortisone and adrenocorticotropic hormone on infection. *Annu. Rev. Med.*, **3**, 1–24.

THOMAS, P.T., BUSSE, W.W., KERKVLIET, N.I., LUSTER, M.I., MUNSON, A.E., MURRAY, M., *et al.* (1990) Immunologic effects of pesticides. In: *The Effects of Pesticides on Health*, Vol. XVIII. pp. 259–295. New York: Princeton Scientific.

VAN VLEM, B., VANHOLDER, R., DE PAEPE, P., VOGELAERS, D. and RINGOIR, S. (1996) Immuno-modulating effects of antibiotics: literature review. *Infection*, **24**, 275–291.

VERDIER, F., VIRAT, M., SCHWEINFURTH, H. and DESCOTES, J. (1991) Immunotoxicity of bis(tri-n-butyltin)oxide in the rat. *J. Toxicol. Environ. Health*, **32**, 307–317.

VIAL, T., NICOLAS, B. and DESCOTES, J. (1996) Clinical immunotoxicology of pesticides. *J. Toxicol. Environ. Health*, **48**, 215–229.

VLEEMING, W., VAN AMSTERDAM, J.G.C., STRICKER, B.H.C. and DE WILDT, D.J. (1998) ACE inhibitor-induced angioedema. Incidence, prevention and management. *Drug Safety*, **18**, 171–188.

VOS, J.G., KRAJNC, E.I., BEEKHOF, P.K. and VAN LOGTEN, M.J. (1983) Methods for testing immune effects of toxic chemicals: evaluation of the immunotoxicity of various pesticides in the rat. In: *IUPAC Pesticide Chemistry: Human Welfare and the Environment* (Miyamoto, T., ed.), pp. 497–504. Oxford: Pergamon Press.

WATSON, R.R. (1990) *Drugs of Abuse and Immune Function*. Boca Raton: CRC Press.

WATSON, R.R. (1993) *Alcohol, Drugs of Abuse and Immunomodulation*. Oxford: Pergamon Press.

WONG, O. (1996) A critical assessment of the relationship between silicone breast implants and connective tissue diseases. *Regul. Toxicol. Pharmacol.*, **23**, 74–85.

ZELIKOFF, J.T. and THOMAS, P.T. (1998) *Immunotoxicology of Environmental and Occupational Metals*. London: Taylor & Francis.

Immunotoxicity Evaluation

8

Guidelines

Because immunotoxicology is still today a relatively new area of toxicology, despite marked evolution in the 1980s and 1990s, it is not surprising that regulatory aspects have been limited, if not largely lacking, until very recently. This chapter is an attempt to cover chronologically major immunotoxicity guidelines up to 1 January 1998. Guidelines on the non-clinical prediction of protein allergenicity have not be included.

US Environmental Protection Agency

In the early 1980s, the US National Toxicology Program (NTP) funded a far-reaching programme in an attempt to define, rationalise, standardise and validate animal models to be used for the non-clinical prediction of the unexpected immunosuppressive effects of xenobiotics.

The selected protocol (Luster *et al.*, 1988) was based on the previously proposed concept of a two-tiered protocol in order to take into account the well-recognised complexity of the immune response and offer a cost-effective approach. B6C3F1 mice, which the NTP had long used for carcinogenicity studies, were selected because the use of a hybrid strain (B6C3F1 mice derive from C57Bl/6N female and DNA/2N male mice) was felt to make immunological assays easier to be performed than an outbred strain due to genetic variability. B6C3F1 mice were treated orally for 14 consecutive days, and initially the interlaboratory validation study included the five following compounds: diethylstilboestrol, benzo[a]pyrene, cadmium chloride, cyclophosphamide and dimethylnitrosamine, which were tested in three different laboratories. This study was later expanded to include over 50 compounds.

The most significant results of this interlaboratory validation study (Luster *et al.*, 1992; 1994) were first the demonstration that interlaboratory reproducibility can be reasonably achieved despite variability due to the use of poorly standardised laboratory reagents, and second, the identification of both statistically and biologically significant correlations between functional immune assays and host resistance models. The plaque forming cell assay was found to be the best predictor of immunosuppression, and a 100 per cent prediction could be achieved when combining plaque forming cell assay and NK cell activity. Other reliable predictors were found to be lymphocyte subset analysis and cell-mediated immunity, either lymphocyte proliferation assay or delayed-type hypersensitivity.

The Office of Pesticide Programs of the US Environmental Protection Agency issued a guideline in 1982 for the immunotoxicity evaluation of pesticides (Subdivision M – Pesticide Assessment Guidelines) requiring that any new pesticide is evaluated using assays to be included in a tier-1 and a tier-2 as listed in Table 8.1. Because this guideline was undoubtedly published very prematurely, when neither standardised nor validated assays had been adequately identified, it soon became obvious that modifications and upgrading were required. In 1988, the US National Agricultural Chemicals Association proposed that the immunotoxicological evaluation of pesticides is performed during conventional 90-day rat studies (tier-1), while a specific immunotoxicity 28-day rat study would only be performed when necessary. The final decision regarding the immunotoxicity of the compound would be based on the results of host resistance assays included in a third tier.

The same year, the US EPA proposed a first draft revision of the initial guideline (Sjoblad, 1988). Mice and rats remained the preferred species, but other species were accepted provided their use could be documented. Healthy adult animals of one sex could be used, with at least ten animals per group. The sensitivity of each selected assay would have to be confirmed by using a positive control group of at least five animals, and an additional group of 20 animals would have to be used for assessing reversibility. At least three dose levels would have to be used, with the lower dose inducing no immunotoxic effect, and the higher dose no lethality or weight loss. The duration of exposure would have to be at least 30 days. Animals would have to be checked clinically daily, and clinical chemical as well as postmortem examinations included. Assays were still divided into two different tiers. The assays of the first tier would include total serum immunoglobulin levels or direct plaque forming cell assay to assess humoral immunity; mixed leukocyte reaction, delayed-type reaction, or T lymphocyte cytoxicity to assess cell-mediated immunity; and NK cell activity or macrophage functions to assess non-specific defences. Tier-2 would have to be performed only when positive or uninterpretable results are obtained in tier-1, or when other data, such as a chemical structure close to that of a known immunotoxicant, are available. Assays included in the second tier would have to evaluate the time course of recovery from immunotoxicity, the resistance to experimental infections, such as infection to *Listeria monocytogenes* or herpes simplex virus. Additional tests would have to be considered, such as serum complement levels, T-independent antibody response, or lymphocyte subset analysis.

Finally, immunotoxicity testing guidelines for pesticides under the EPA's Toxic Substances Contact Act (TSCA) were recently updated in final form (*Federal Register*, 1997). Although allergy and autoimmunity were included in the definition of immunotoxicity, the scope of this guideline remained restricted to immunosuppression. Interestingly, emphasis was put on functional assays, even though it was clearly stated that the proposed assays did not represent a comprehensive assessment of immune function, and relevant information on pathological changes in lymphoid organs were considered to be obtained in the course of conventional toxicity testing. Mice and rats were the recommended species, but one species can be used, unless no adequate ADME data are available. At least eight animals of each sex per group should be used, unless one sex is known to be more sensitive to the immunotoxicant under scrutiny. Animals have to be maintained in controlled conditions and given one of several doses in order to produce a dose–response

Table 8.1   Tier-1 and tier-2 assays to be included for the immunotoxicity evaluation of pesticides according to US EPA Subdivision M – Pesticide Assessment Guidelines (1982)

TIER I

Immunopathology
  Haematology (blood cell counts)
  Total, spleen, thymus and liver weight
  Spleen cellularity
  Histology of the spleen, thymus and lymph nodes
Humoral immunity
  Plaque forming cell assay (primary response to sheep erythrocytes)
Cell-mediated immunity
  Mitogen-induced lymphocyte proliferation and mixed leukocyte reaction
Non-specific immunity
  Natural killer (NK) cell activity

TIER II

Immunopathology
  Enumeration of B and T cells
Humoral immunity
  Plaque forming cell assay (secondary response to sheep erythrocytes)
Cell-mediated immunity
  T lymphocyte cytotoxicity
  Delayed-type hypersensitivity
Non-specific immunity
  Enumeration and phagocytic activity of peritoneal macrophages
Resistance assays
  Implantation of syngeneic tumour cells (PYB6 fibrosarcoma, B6F10 melanoma)
  Bacterial infections (*Listeria monocytogenes*, streptococci)
  Viral infections (influenza)
  Parasitic infections (*Plasmodium yoleii*)

---

relationship, and a no-observed immunotoxic effect level. Ideally, the lowest dose should not induce immunotoxic effects, and the highest dose should produce measurable, but sign(s) of moderate general toxicity, such as a 10 per cent loss in body weight, but not stress, malnutrition or fatalities. Unsurprisingly, the key functional assay is the plaque forming cell (PFC) assay to be performed in animals dosed for at least 28 days, to take into account the half-life of immunoglobulins in rodents. However, immunoglobulin quantification by ELISA is also an accepted alternative to the PFC assay. In the event that a significant suppression of PFC response is observed, phenotypic lymphocyte subset analysis using flow cytometry and NK cell activity should be considered. Finally, a positive control group of at least eight animals, treated with cyclophosphamide should be used. It is unclear why immunotoxicity studies should include a positive control group and conventional repeated dose administration studies should not. This latter aspect is unlikely to be readily acceptable in most European countries committed to restricting the use of laboratory animals included in toxicity evaluation studies.

After 15 years of discussion, the initial EPA immunotoxicity testing guidelines for pesticides has therefore come to maturity with a relatively balanced approach, including histological and functional endpoints. It however remains to be seen whether and to what extent these guidelines will be accepted and implemented by the industry and regulatory agencies.

Council of the European Communities

The Council of the European Communities was the first regulatory agency to issue a recommendation emphasising the need for an immunotoxicological evaluation of new medicinal products (Council of the European Communities, 1983). Surprisingly, this text was paid very limited, if any, attention by pharmaceutical companies and national regulatory agencies within Europe and has never been enforced by national regulatory agencies in Europe, or supported by lobbying groups.

This recommendation states that because of the development of immunology and its acknowledged importance, the interferences of medicinal products with the immune system should be considered, even though such interferences are not expected from the intended therapeutic use of these products, as they may result in potentially severe adverse reactions, such as infectious diseases and neoplasias. Emphasis was given to the histological examination of the spleen, thymus and lymph nodes, in the hope that observed changes could be suggestive of immunotoxicity. If such changes were noted, additional assays would have to be performed, although no indications on suggested assays were actually provided.

Several comments can be proposed:

- This recommendation, whatever its relevance, restricts the scope of immunotoxicology to immunosuppression only. Historical reasons, as depicted in the first chapter of this volume, are likely to explain this restriction, as immunosuppression was undoubtedly the major, if not only, immunotoxicological concern in the early 1980s. Unfortunately, considerations of hypersensitivity and autoimmunity are totally missing in this recommendation.

- Histology was considered the key endpoint to identifying unexpectedly immunosuppressive medicinal products. Even though macroscopic and microscopic examinations of the thymus, spleen and main lymph nodes can probably provide helpful information, they are unlikely to ensure that every immunotoxicant can be detected using histological changes only as indicators of immunotoxicity. It was indeed shown that thymic atrophy can be seen prior to any other immunotoxic effect in rats exposed to organotins, such as tributyltin oxide (Verdier *et al.*, 1991), but depression of functional immune endpoints can also be seen prior to thymic atrophy in rats treated with cyclosporin. Nevertheless, more recent texts, such as OECD guideline 407, similarly gave preference to histological examination versus immune function assays in contrast to the last requirement of EPA in the context of the Toxic Substances Contact Act.

- This recommendation is clearly a very careful, or overprudent text, in that no requirement is made regarding the most reliable immunological endpoints or assays to be used. Even though the 'cookbook approach' of toxicology, once largely favoured, is nowadays more and more regarded as obsolete or inappropriate for sound safety evaluation, the lack of any detailed requirement is indicative of uncertainties from the legislator's perspective regarding non-clinical immunotoxicity evaluation due to the lack of standardised and validated assays at that time, and this might well explain why this recommendation has never actually been enforced.

Whatever its limitations, this recommendation was the first to be issued regarding the non-clinical immunotoxicity evaluation of new medicinal products. More than 15 years later, it is still unclear why it failed to have any consequences on the regulatory assessment of new medicinal products as well as on the industry approach of new products. Even though it is possible today to predict whether a new chemical entity can exert unexpected immunosuppressive effects, very little has been done in the past 10 or 15 years to implement the non-clinical immunotoxicity evaluation of new medicinal products, despite many meetings and workshops devoted to immunotoxicology, as well as review and original papers on this topic, during this period of time. That a comprehensive set of guidelines could be recently released regarding the non-clinical immunotoxicity evaluation of pesticides demonstrates that this issue is timely, and it is therefore unclear why this should be a timely issue for pesticides only, and not for other major types of chemicals to which human beings are exposed, such as medicinal products.

It is unsure, and probably unlikely, that conventional non-clinical toxicity studies can reliably detect whether a new chemical entity is able to exert immunosuppressive effects with opportunistic infections and lymphomas are the main clinical consequences. For example, most, if not all, AIDS drugs in use today have been approved without a comprehensive study of their possible adverse influences on the rodent and human immune system, although they are intended for specific use in immunocompromised patients. If a new non-immune-target medicinal product was to cause lymphomas and opportunistic infections in a significant fraction of the population, how could the pharmaceutical company who marketed the product convincingly explain why this aspect of safety was ignored, as one major international regulatory agency in the world officially stressed the need for such an evaluation, years ago? Anyway, for the time being, the non-clinical immunotoxicity evaluation of new medicinal products remains a perspective (De Waal *et al.*, 1996).

The Japanese authorities' drug antigenicity requirement

Unofficial guidelines from the Japanese health authorities exist regarding the antigenicity of medicinal products, even though the last draft version is markedly down-sized as compared to the initial version.

Whatever their molecular weight, new medicinal products should be evaluated using the raw material and a protein-conjugate in three different assays (Udaka, 1992):

- Guinea-pig anaphylactic shock with two sensitising oral or transcutaneous injections at one-week intervals, and intravenous or intradermal challenge after ten days.
- Passive cutaneous anaphylaxis reaction in rats.
- Indirect haemagglutination technique in mice with two subcutaneous sensitisation injections at 2–4 week intervals.

Unfortunately, this protocol, whatever the theoretical logic which can be found in it, has never been standardised or validated. Use of this protocol is therefore very likely to generate falsely reassuring data, as small-molecular-weight substances with limited, if no, intrinsic chemical reactivity (as is the case for the vast majority of medicinal products) cannot be reasonably expected to result in reliably positive findings. The question remains why the international harmonisation (ICH) process did not address this very debated, and highly debatable, issue.

OECD guideline 407

In 1996, the Organization for Economic Development and Cooperation (OECD) released a revised version of guideline 407, adopted on 27 July 1995. This guideline comprises the basic repeated dose toxicity study that may be used for chemicals on which a 90-day study is not warranted (Koeter, 1995). The duration of exposure is normally 28 days. This revised guideline put more emphasis on neurological and immunological effects. However, emphasis as regards immunotoxicity is placed only on the gross necropsy and standard histological examination of the lymphoid organs in the control and high dose groups, namely the thymus, spleen, lymph nodes (including one lymph node covering the route of administration and one distant lymph node to cover systemic effetcs), and Peyer's patches.

Both positive and negative aspects can be found in the revised version of OECD guideline 407. The first and major positive aspect is the requirement that immunotoxicity endpoints are to be systematically included in the non-clinical evaluation of every new chemical. Xenobiotics, especially non-medicinal products, which do not warrant a 90-day repeated administration study, are at least tested in a 14- or more often 28-day rat study and such studies should include at least the histological examination of the main lymphoid organs.

The major negative aspect of this guideline is the limited number of selected immunotoxicity endpoints. Most immunotoxicologists think that histological endpoints cannot reliably predict whether a new chemical is likely to exert immunosuppressive effects. There is a general agreement among immunotoxicologists that functional endpoints of the immune response are absolutely needed. Therefore, even though OECD guideline 407 is strictly adhered to, it is not possible to ensure that no unexpected and severe immunotoxicity can occur after a new chemical entity is introduced onto the market. Hopefully, ongoing discussions regarding OECD guideline 408 dealing with 90-day repeated administration studies might come to the decision to include immune function endpoints. In addition, the newly released US EPA guidelines on the non-clinical immunotoxicity evaluation of pesticides further support the view that additional endpoints are required to ensure that a more appropriate non-clinical evaluation of immunotoxicity is achieved.

The US FDA Red Book II

In consideration of the need for defining immunotoxicity with regard to regulated products of the US Food and Drug Administration, an intra-agency taskforce was convened to arrive at a definition of immunotoxicity. Pursuant to these efforts, immunotoxicity testing guidelines in draft form (Hinton, 1992) were proposed to be included in a second edition of *Toxicological Principles for the Safety Assessment of Direct Food and Color Additives Used in Food* (often referred to as the Red Book).

Two levels of immunotoxicity testing are defined, namely level I, which does not require injection of an antigen to animals, so that selected endpoints can be measured in the same animals as those used in the standard toxicity studies, and level II, which is defined as functional testing and requires a concurrent satellite group of animals. Primary indicators of immunotoxicity include total and differential white blood cell counts, clinical chemistry screen endpoints (such as liver enzymes), lymphoid organ weights, and standard histological examination of the spleen, thymus, lymph nodes and Peyer's patches. Expanded level I serum chemistry screen might include quantification of serum immunoglobulin and auto-antibody levels, NK cell and macrophage functional activity, assessment of haemolytic complement, and enhanced histopathology, based on the immunostaining of lymphocytes in tissues. Level II testing is suggested to include delayed-type hypersensitity responses (such as contact hypersensivitity to oxazolone or dinitrochlorobenzene) and humoral responses to T-dependent and T-independent antigens.

In fact, despite several revisions, a final version of these immunotoxicity guidelines has not yet been published, presumably because they do not rely on a cost-effective assessment of issues associated with the possible immunotoxicity of food products and food additives.

Other guidelines

Immunotoxicology was included in several guidelines in final or draft form, such as the European guidelines on biotechnology-derived products, the ICH document on biologicals, the French guidelines on veterinary medicines, and the US FDA immunotoxicity testing framework for medical devices. In general though, these guidelines only indicate the need for immunotoxicity endpoints to be included in the safety evaluation of these compounds, but no clear indication is provided on which assays should be performed.

Perspectives

Although regulatory aspects have long been (Descotes, 1986) and still are (De Waal *et al.*, 1996; Harling, 1996) a matter of concern for immunotoxicologists, a limited number of guidelines are available, as tentatively shown in this chapter.

Several issues and needs can be identified (Descotes, 1998). At least four issues are proposed to be essential:

- The first issue is whether hazards related to immunotoxic effects have been adequately identified and may pose a threat to human health. As discussed previously in this volume, there is a large body of evidence that immunotoxic effects can result in a variety of potentially severe adverse consequences to human health.

- The second issue is whether and to what extent such immunotoxic hazards can affect human beings in given conditions of exposure. Several risks have been adequately characterised, but others remain excessively uncertain. For instance, immunosuppression, when severe, can result in infectious complications and lymphomas, but fewer data suggest that immunodepression is associated with similar, but less severe and less frequent adverse events. The clinical experience gained with therapeutic cytokines showed, to some extent, a parallelism between the immunostimulating potency of medicinal products and the frequency and severity of adverse events. Hypersensitivity reactions are relatively common. Finally, auto-immune reactions induced by xenobiotics, although seemingly very rare, are usually severe.

- The third issue is the acceptability of immunotoxic risks. There is no definition of what an acceptable risk is. In any case, several immunotoxic effects, such as infectious complications and lymphomas associated with immunosuppression, acute cytokine syndromes, anaphylactic shocks and systemic autoimmune reactions, can be severe and sometimes life-threatening.

- If they are judged unacceptable, the next issue is whether reasonably adequate methods are available to assess immunotoxic risks. Extensive efforts have been paid to the design and evaluation of methods for predicting the unexpected immunosuppressive effects of xenobiotics. In contrast, other aspects of immunotoxicity, such as unexpected immunostimulation, autoimmune reactions and hypersensitivity, received only limited attention until recently.

Several unmet needs might explain why adequately standardised and validated methods are not available to evaluate every potential immunotoxic effect:

• Immunotoxicity should not be restricted to immunosuppression. Despite the fact that hypersensitivity is the commonest immunotoxic risk of drug treatment and occupational exposure to industrial chemicals, immunosuppression was the primary focus of immunotoxicological investigations. The time has come to focus on other critical aspects of immunotoxicity.

• Although science-oriented studies are absolutely essential, efforts should also be paid to conduct less science-oriented immunotoxicological studies, for instance to establish the predictive value of available methods using less prototypic compounds than the immunosuppressant cyclosporin, the contact sensitiser dinitrochlorobenzene or the respiratory allergen isothiocyanate.

• The development of clinical immunotoxicology is another major need. It is often unknown how immunotoxic findings in rodents compare with immunotoxic effects in humans. As available biomarkers of immunotoxicity are inadequate in most instances, efforts should be paid to designing better endpoints.

• The majority of methods and assays for use in immunotoxicity evaluation derive from immunology. Even though assays, such as the plaque assay, are reliable indicators of immunosuppression, models designed to address immunotoxicity issues specifically are needed.

• Regulations in immunotoxicology can have a major impetus, as laboratories throughout the world would have to generate data and thus gain experience to design new methods and concepts.

References

COUNCIL OF THE EUROPEAN COMMUNITIES (1983) *Official Journal of the European Communities*, no. L332/11, 26 October 1983. Brussels.

DESCOTES, J. (1986) Immunotoxicology: health aspects and regulatory issues. *Trends Pharmacol. Sci.*, **7**, 1–3.

DESCOTES, J. (1998) Regulating immunotoxicity evaluation: issues and needs. *Arch. Toxicol.*, **20**, 5293–5299.

DE WAAL, E.J., VAN DER LAAN, J.W. and VAN LOVEREN, H. (1996) Immunotoxicity of pharmaceuticals: a regulatory perspective. *Toxicol. Ecotoxicol. News*, **3**, 165–172.

Federal Register (1997) Toxic Substances Control Act Test Guidelines. Final Rule. **62**, 43819–43864.

HARLING, R.J. (1996) Perspectives on immunotoxicity regulatory guidelines. *Inflamm. Res.*, **45**, S69–S73.

HINTON, D.M. (1992) Testing guidelines for evaluation of the immunotoxic potential of direct additives. *CRC Rev. Food Sci. Nutr.*, **32**, 173–190.

KOETER, H.B.W.M. (1995) International harmonization of immunotoxicity testing. *Hum. Exp. Toxicol.*, **14**, 151–154.

LUSTER, M.I., MUNSON, A.E., THOMAS, P.T., HOLSAPPLE, M.P., FENTERS, J.D., WHITE, K.L., *et al.* (1988) Development of a testing battery to assess chemical-induced immunotoxicity. National Toxicology Program's criteria for immunotoxicity evaluation in mice. *Fund. Appl. Toxicol.*, **10**, 2–19.

LUSTER, M.I., PORTIER, C. PAIT, D.G., WHITE, K.L., GENNINGS, C., MUNSON, A.E. and ROSENTHAL, G.J. (1992) Risk assessment in immunotoxicology. I. Sensitivity and predictability of immune tests. *Fund. Appl. Toxicol.*, **18**, 200–210.

LUSTER, M.I., PORTIER, C., PAIT, D.G., ROSENTHAL, G.J., GERMOLEC, D.R., CORSINI, E., *et al.* (1994) Risk assessment in immunotoxicology. II. Relationships between immune and host resistance tests. *Fund. Appl. Toxicol.*, **21**, 71–82.

SJOBLAD, R.D. (1988) Potential future requirements for immunotoxicology testing of pesticides. *Toxicol. Indust. Health* **4**, 391–395.

UDAKA, K. (1992) Cellular and humoral mechanisms of immunotoxicological tissue manifestations induced by immunotoxic drugs. In: *Toxicology from Discovery and Experimentation to the Human Perspective* (CHAMBERS, P.L., CHAMBERS, C.M., BOLT, H.M. and PREZIOSI, P. eds), pp. 93–100. Amsterdam: Elsevier.

VERDIER, F., VIRAT, M., SCHWEINFURTH, H. and DESCOTES, J. (1991) Immunotoxicity of bis(tri-n-butyltin)oxide in the rat. *J. Toxicol. Environ. Health*, **32**, 307–317.

9

Histopathology

Histological examination is an obligatory component of non-clinical toxicity evaluation as histological changes are a primary effect of toxic exposures. These changes may be expressed as quantitative or qualitative alterations in the normal structure and function of target organs. But, as histopathological diagnosis is qualitative and often represents a subjective judgement on the nature and the expected consequence of a specific lesion, a central question is to what extent histopathology can be effectively used in the non-clinical immunotoxicity evaluation of xenobiotics. Because routine histopathology is necessarily a component of conventional toxicity testing and no additional animals are required, it is often considered to be an essential component of non-clinical immunotoxicity evaluation.

The pathologist can identify altered architecture of the lymphoid organs and changes in the distribution of various cell populations, which are not readily discernible using immune function tests. Therefore, routine histopathology has been proposed as a straightforward and cost-effective approach to identify potentially immunotoxic compounds (Basketter *et al.*, 1995; Bloom *et al.*, 1987; Schuurman *et al.*, 1994). Nevertheless, a number of investigators consider immune function tests to be more sensitive than histopathological examination. Potent immunotoxicants indeed produce frank histopathological changes, which can be easily detected, even in the course of short-term toxicity studies, whereas less potent compounds produce no morphological changes in similar experimental conditions. In addition, depending on the mechanism of action, even potent immunosuppressive drugs, such as cyclosporin, may not induce marked histological changes (Ryffel *et al.*, 1983), except when *ad hoc* histological examination of lymphoid organs, such as the thymus (Beschorner *et al.*, 1987), is performed.

Anyway, a standard histological examination of the major lymphoid organs, namely the thymus, spleen, bone marrow, lymph nodes and Peyer's patches, as part of the routine toxicity screen, is often considered to be an early, if reliable, indicator of possible immunotoxicity, a view supported, not without debate, by the recently updated OECD guideline 407 (see Chapter 8).

It is also common practice to include non-histological, non-functional endpoints in the histopathology package of non-clinical immunotoxicity evaluation, such as clinical chemistry, serum immunoglobulin levels and leukocyte subset analysis.

Organisation of the lymphoid tissue

Unlike the other organ systems, the cellular components of the immune system are widely disseminated throughout the body (Schuurman *et al.*, 1991). Cells of the immune system all arise from multipotent stem cells, which are found in the liver during foetal development and later in life in the bone marrow. Multipotent stem cells follow various pathways of differentiation to give rise to red blood cells (erythrocytes), macrophages and polymorphonuclear leukocytes, platelets, and lymphocytes. Immunocompetent cells are disseminated in lymphoid organs classified into primary and secondary lymphoid organs, whether the proliferation and differentiation of immunocompetent cells is antigen-dependent or not. The primary (or central) lymphoid organs include the thymus (the bursa of Fabricius in birds) and the bone marrow, while the spleen, lymph nodes and specialised lymphoid tissues, such as Peyer's patches, are secondary (or peripheral) lymphoid organs.

Histopathology of the lymphoid tissue

Key practical aspects of the histological examination of lymphoid organs, and the interpretation of results have been reviewed (Gopinath, 1996; Lebish *et al.*, 1986; Schuurman *et al.*, 1991, 1994).

Routine histopathological examination

Lymphoid organ weight

Organs are typically weighed prior to routine histology examination following completion of repeated administration toxicity studies. It is generally recommended that weighing major lymphoid organs, namely the thymus, spleen and selected lymph nodes, is to be included in the initial screening for potential immunotoxicity.

Selection of relevant lymph nodes can be as follows: draining lymph nodes should be selected according to the route of entry, e.g. mesenteric lymph nodes for the oral route, or bronchial lymph nodes for inhalation, whereas dormant lymph nodes located at a distance from the route of entry are examined to detect possible systemic effects. The popliteal lymph node is often considered the most suitable dormant lymph node. Because of anatomical variability, possible adherence to non-lymphoid fatty tissue, and involution of lymphoid organs, such as the thymus, great care should be exercised when removing lymph nodes. The determination of cellularity is often considered a useful addition to the weighing of lymphoid organs.

Standard histological examination of lymphoid organs

The tissues and organs to be examined must be fixed as quickly as possible after necropsy. Several fixatives allow conventional and immunochemistry staining. Formalin fixation, paraffin embedding and staining with haematoxylin and eosin, are often the recommended procedures to identify changes in the architecture of lymphoid tissues and the morphology of most immunocompetent cells.

Interpretation of observed histological changes

Because the immune system is a complex and dynamic system involving many interactions between soluble components and immunocompetent cells under the influence of a variety of external factors, the histology of lymphoid organs is very variable (Gopinath, 1996; Schuurman *et al.*, 1994).

Major external sources of variability include stress, steroid hormones, antigenic load, nutritional status and age. Psychological stress, disease-mediated stress and stress induced by severe systemic toxicity all have a marked effect on the immune system, essentially via the release of the glucocorticosteroid hormones. Lymphopenia is seen almost immediately and shrinkage of the thymic cortex may occur within days. There is a general agreement that animals of the high dose group in non-clinical immunotoxicity evaluation studies should not develop marked systemic toxicity in order to avoid stress-mediated histological and immunological changes, which will make the interpretation of findings difficult.

The status of the lymphoid organs is under the balanced influence of hormones, particularly oestrogens and androgens. The thymus is again the primary target when this exquisite balance is shifted to either side.

The antigenic load is considered to have a marked influence on the architecture of lymphoid organs. This is the reason why it is recommended to select both dormant and activated lymph nodes according to the route of entry. Obviously, housing conditions, food and the microbiological status of animals are all likely to influence the histology of lymphoid organs. Nutrition is another critical factor, as undernutrition, which is often the hallmark of systemic toxicity, can result in thymus atrophy.

Finally, age affects lymphoid organ histology, with thymic involution as the most notable finding. Obviously, the role of these confounding factors should be carefully investigated when interpreting the results of histological examination of lymphoid organs to avoid erroneous or misleading conclusions.

The thymus is a major target organ of immunotoxicity and a number of immunotoxicants, such as cyclosporine, organotins or dioxin, have been shown to induce thymic atrophy in rodents (Kendall and Ritter, 1991). However, every lymphoid organ can be affected. Depletion of lymphoid organs in immunocompetent cells is usually a consequence of immunosuppression, whereas expansion and proliferation of various cell types and areas, such as lymphoid follicles, and germinal centres, either in lymph nodes or in the thymus, are typical consequences of immunostimulation. Hypersensitivity and autoimmune processes result in histological changes affecting specific components of the lymphoid system depending on the pathophysiological process.

Immunohistological examination

Because standard histological examination is not generally accepted as a totally reliable method to predict immunotoxicity, particularly unexpected immunosuppression, other techniques have been proposed or designed to enhance the predictibility of histopathology in the context of immunotoxicity evaluation. The term 'advanced' or 'enhanced' pathology is often used to refer to these techniques.

Immunohistology and immunohistochemistry are used for the phenotyping of immunocompetent cells using specific antibodies against cell membrane markers, and the study of cytoplasmic components and immune complexes (Muro-Cacho *et al.*, 1994). The freezing of tissue sections is the recommended procedure to preserve the architecture and antigenicity of cell surface markers. A variety of immunohistochemical techniques have been developed using enzymatic detection reactions or fluorochromes as labelling substances. With these techniques, it is possible to achieve immunostaining for surface markers, specific enzymes, immunoglobulins or cytokines and thus to provide information on the functional status of the immune system, at least to some extent.

Other advanced techniques include *in situ* hybridation, and quantitative techniques using microscope image analysis. However, the value of these advanced histopathological tools, as adjuncts to conventional histological examination to improve the detection of potential immunotoxicants, remains to be established.

Non-histological, non-functional endpoints

Several non-histological, non-functional endpoints are commonly included in what used to be called the histopathological evaluation of immunotoxicants. These endpoints can be divided into three categories: clinical chemistry, serum immunoglobulin levels and leukocyte analysis.

Clinical chemistry

A number of clinical chemistry parameters are routinely included in conventional toxicity testing. There is no reason to suggest these parameters should not be included in non-clinical immunotoxicity studies as they can give clues to understanding observed changes. In addition, it could be interesting to include specific or supposedly specific clinical chemistry parameters, such as inflammation proteins. However, this has so far seldom been considered. Similarly, current advances in the assays of cytokines could lead to better acceptance of such assays as part of clinical chemistry parameters. Cytokine assays are dealt with in Chapter 11.

Serum immunoglobulin levels

Serum immunoglobulin levels have often been reported to be altered following exposure to medicinal products and chemicals. It is indeed logical to assume that changes in serum immunoglobulin levels are indicative of alterations in humoral immunity as serum immunoglobulin levels depend on the humoral arm of the immune responsiveness. However, observed changes in serum immunoglobulin levels may not always correlate with the influence of a given immunotoxicant on antigen-specific antibody responses. The measurement of serum immunoglobulin levels is not a functional assay and therefore lacks the sensitivity of functional assays.

In most instances, only IgM and IgG serum levels are measured, but measuring IgA and IgE serum levels as well, may be recommended due to the physiological importance of these immunoglobulins in mucosal immunity and anaphylaxis, respectively. The measurement of serum immunoglobulin levels cannot provide useful information after short-term exposure. No reduction in serum immunoglobulin levels due to decreased synthesis induced by an immunotoxic exposure can indeed be seen before immunoglobulins are normally metabolised by the body. This is exemplified by the negative results of the interlaboratory validation study conducted under the auspices of the US National Toxicology Program, in which mice were exposed for only 14 consecutive days, a period obviously too short to detect an effect on serum immunoglobulin levels.

Various methods can theoretically be used to measure serum immunoglobulin levels. Besides traditional methods, such as radial immunodiffusion and immuno-electrophoresis, new methods, particularly ELISA (Law, 1996), are increasingly recommended:

- *Radial immunodiffusion* is a simple technique to measure a small amount of antigen in a mixture of several antigens, by using a specific antibody. The monospecific antiserum is incorporated in a thin gel layer, which is deposited on a glass plate and the mixture of antigens is placed into a well in the gel. The binding of the antigen and the corresponding antibody generates a precipitation area around the well, the external diameter of which is proportional to the antibody concentration. The antigen concentration is measured by the use of standard curves established from known antigen concentrations. Large amounts of sera are not required. However, this is a relatively slow technique with a detection threshhold of only 1.5–5 µg/ml, which is however adequate to assay serum IgG and IgM levels.

- *Immuno-electrophoresis* was designed to take advantage of electophoretic mobility and antigenic specificity. A mixture of antigens is introduced into the gel medium (agarose) and an electric current is applied to separate antigenic molecules according to electrophoretic mobility. A polyspecific antiserum is then placed to diffuse perpendicularly to the electric current. The free diffusion of antigens and antibodies towards each other induces precipitation bows. The interpretation of results is hampered by the semiquantitative evaluation in comparison to pool control sera.

- *ELISA* is by far the preferred technique today. ELISAs (Enzyme-Linked Immunosorbent Assay) are more sensitive, easily automated and applicable to the measurement of all classes of immunoglobulins. In ELISA, an enzyme label is linked either to the antigen or the antibody. Many different enzymes have been used as tracers in ELISAs, such as horseradish peroxidase and β-galactosidase. In the immobilised antigen technique, a hapten-conjugate is prepared to be absorbed onto the surface of a solid phase. The solid phase is later incubated with the specific antiserum together with standards and samples containing the analyte. The amount of antibody bound to the immobilised antigen bound to the plate (solid phase) at equilibrium is inversely related to the concentration of added analyte. It is detected using a second enzyme-labelled antibody and a calibration curve is obtained to measure the analyte concentration.

- Other possible, but seldom used methods include electroimmunodiffusion, immunonephelemetry and radioimmunoassays. Electroimmunodiffusion or rocket immunodiffusion is based on the principle of radial immunodiffusion, but the diffusion time is reduced because antigen diffusion into gel is accelerated by an electric current. Compared to radial immunodiffusion, electroimmunodiffusion is quicker and more sensitive (approximately 1 µg/ml), but also more complex. Immunonephelemetry is based on the physical laws of light diffusion by particles in suspension. The precipitation of antibody–antigen complexes induces changes in light diffusion which can be measured. The addition of a monospecific serum allows measurement of the corresponding immunoglobulins. This is a technique easy to perform, reproducible and easily automatised. The detection threshold is approximately 1 µg/ml. Radioimmunoassays are seldom used to measure total serum immunoglobulins, except IgE.

Leukocyte analysis

As leukocytes comprise the majority of immunocompetent cells, leukocyte analysis is an essential component of immunotoxicity evaluation.

White blood cell counts

At the simplest level, data on white blood cell counts should be available at different times during exposure. White blood cells (leukocytes) can be counted using automated, semi-automated or manual methods (Suber, 1994). Automated or semi-automated methods, such as aperture impedance and laser light scatter methods, reduce the variability due to human errors when manual methods, such as the haemocytometer method, are used.

There are several interspecies differences. The morphology of neutrophils is somewhat different in humans, rodents and dogs. In rats, the neutrophil percentage is very small at birth and increases with age. Lymphocytes are the predominant leukocytes in mice and young rats. Basophils are seldom found in the circulation of mice, rats and dogs. Fischer 344 rats are prone to developing leukemia.

Lymphocyte surface marker analysis

The major aspect is lymphocyte subset analysis. The determination of phenotypic markers on immunocompetent cells using fluorescent activated cell sorter (FACS) and fluorescent cell counter is increasingly performed in the context of non-clinical and clinical immunotoxicity evaluation (Ladics *et al.*, 1994, 1997). Importantly though, lymphocyte subset analysis should not be regarded as a functional assay. Nevertheless, its usefulness for predicting immunosuppression has been clearly demonstrated by the NTP interlaboratory validation study in B6C3F1 mice as well as in several other interlaboratory validation rat studies.

A major problem today with lymphocyte subset analysis is the selection of markers. Only a very small number of markers, essentially markers of B and T cells as well as T $CD4^+$ and T $CD8^+$ lymphocytes have been studied with many immunosuppressive compounds so that the usefulness of new or recently proposed markers remains to be established for immunotoxicology evaluation. Nevertheless, markers of more specific lymphocyte subsets are expected to be increasingly used (Burchiel *et al.*, 1997).

Another issue is the selection of species. Most studies have been conducted in mice due to the limited availability of reagents in the rat. However, recent studies largely confirmed the usefulness of lymphocyte surface markers in the rat (Ladics *et al.*, 1994, 1997). In contrast, only limited information is available in monkeys (Verdier *et al.*, 1995), and still more so in dogs.

Other leukocyte analytical techniques

Because leukocytes are so important in the immune response, the enumeration of leukocytes at various sites in addition to peripheral blood or spleen can prove helpful for immunotoxicity evaluation.

Mononuclear phagocytes are pivotal in the clearance of inhaled particles from alveoli of the lung. The enumeration of these cells in bronchoalveolar lavage fluid and the study of their functions can prove useful, but available methods usually require skill and care to provide reliable results, and they are not yet adapted to the conduct of routine immunotoxicology evaluation studies in the full respect of Good Laboratory Practice. The same applies to a large extent to enumeration of mononuclear phagocytes from the peritoneal cavity. Finally, Langerhans cells can be enumerated in the skin using immuno-histochemistry or morphometric techniques.

References

BASKETTER, D.A., BREMMER, J.N., BUCKLEY, P., KAMMÜLLER, M.E., KAWABATA, T., KIMBER, I., *et al.* (1995) Pathology considerations for, and subsequent risk assessment of, chemicals identified as immunosuppressive in routine toxicology. *Fd Chem. Toxicol.*, **33**, 239–243.

BESCHORNER, W.E., NAMNOUN, J.D., HESS, A.D., SHINN, C.A. and SANTOS, G.W. (1987) Cyclosporin and the thymus. Immunopathology. *Am. J. Pathol.*, **126**, 487–496.

BLOOM, J.C., THIEM, P.A. and MORGAN, D.G. (1987) The role of conventional pathology and toxicology in evaluating the immunotoxic potential of xenobiotics. *Toxicol. Pathol.*, **15**, 283–292.

BURCHIEL, S.W., KERKVLIET, N.L., GERBERICK, G.F., LAWRENCE, D.H. and LADICS, G.S. (1997) Assessment of immunotoxicity by multiparameter flow cytometry. *Fund. Appl. Toxicol.*, **38**, 38–54.

GOPINATH, C. (1996) Pathology of toxic effects on the immune system. *Inflamm. Res.*, **2**, S74-S78.

KENDALL, M.D. and RITTER, M.A. (1991) *The Thymus in Immunotoxicology*. Chur: Harwood.

LADICS, G.S. and LOVELESS, S.E. (1994) Cell surface marker analysis of splenic lymphocyte populations of the CD rat for use in immunotoxicological studies. *Toxicol. Meth.*, **4**, 77–91.

LADICS, G.S., CHILDS, R., LOVELESS, S.E., FARRIS, G., FLAHERTY, D., GROSS, C. (1997) An interlaboratory evaluation of quantification of rat splenic lymphocyte subtypes using immunofluorescent staining and flow cytometry. *Toxicol. Meth.*, **7**, 99–108.

LAW, B. (1996) *Immunoassay. A Practical Guide*. London: Taylor & Francis.

LEBISH, I.J., HURVITZ, A., LEWIS, R.M., CRAMER, D.V. and KRAKOWKA, S. (1986) Immunopathology of laboratory animals. *Toxicol. Pathol.*, **14**, 129–134.

MURO-CACHO, C.A., EMANCIPATOR, S.N. and LAMM, M.E. (1994) Immunohistology and immunopathology. *Immunol. Allergy Clin. N. Am.*, **14**, 401–423.

RYFFEL, B., DONATSCH, P., MADÖRIN, M., MATTER, B.E., RÜTTIMANN, G., SCHÖN, H., *et al.* (1983) Toxicological evaluation of cyclosporin A. *Arch. Toxicol.*, **53**, 107–141.

SCHUURMAN, H.J., KRAJNC-FRANKEN, M.A.M., KUIPER, C.F., VAN LOVEREN, H. and VOS, J.G. (1991) Immune system. In: *Handbook of Toxicologic Pathology* (Haschek, W.M. and Rousseaux, C.G., eds), pp. 421–487. San Diego: Academic Press.

SCHUURMAN, H.J., KUIPER, C.F. and VOS, J.G. (1994) Histopathology of the immune system as a tool to assess immunotoxicology. *Toxicology*, **86**, 187–212.

SUBER, R.L. (1994) Clinical pathology methods for toxicology. In: *Principles and Methods of Toxicology*, 3rd edition (Hayes, W.A., ed.), pp. 729–766. New York: Raven Press.

VERDIER, F., AUJOULAS, M., CONDEVAUX, F. and DESCOTES, J. (1995) Determination of lymphocyte subsets and cytokine levels in nonhuman primates: cross-reactivity of human reagents. *Toxicology*, **105**, 81–90.

10

Assays of Humoral Immunity

Following the introduction of foreign macromolecules (or haptens bound to carrier proteins), the immune system can react by the differentiation and multiplication of B cells, of which some become antibody-producing plasma cells. Humoral immunity refers to the production of antigen-specific antibodies.

In addition to the measurement of serum immunoglobulin levels, which is usually considered a normal component of the histopathological evaluation package as this is a non-functional assay (see Chapter 9), specific antibody responses towards a given antigen are commonly assessed in non-clinical immunotoxicity evaluation. The major advantage of measuring specific antibody response compared to serum immunoglobulin levels is that humoral immunity is explored functionally in conditions that grossly mimic those of a single antigenic stimulation. However, an important limitation is that additional animals are generally considered to be required because of possible changes caused by antigenic stimulation, even though this view is no longer unanimously held (Ladics et al., 1995).

Typically, humoral immunity can be assessed from two different perspectives: first, the measurement of antigen-specific antibody titres/levels in the sera of exposed animals (or humans), and second the determination of antigen-specific antibody-producing cells. Only a few comparative studies have been performed (Temple et al., 1993), so that the selection of either testing procedure remains largely subjective or based on the prior experience of the investigator. The determination of antigen-specific antibody-producing cells, particularly the plaque-forming cell (PFC) assay, has long been the preferred approach by most investigators and this is undoubtedly today the best validated animal model to predict unexpected immunosuppression associated with drug treatment and chemical exposure. However, because limitations, such as insufficient reproducibility, can be noted in the PFC assay, ELISAs are more and more frequently recommended.

Antigenic stimulation

Antigenic stimulation is performed either at the end or immediately after the completion of drug treatment or chemical exposure. To take into account the half-life of immunoglobulins, the duration of exposure should be at least 21 days and more preferably 28 days in rodents. In most cases, only the primary response is explored.

Primary and secondary antibody responses

Depending on whether the contact with the antigen is a first or subsequent contact, the characteristics of antibody responses are markedly different. Major differences between primary and secondary antibody responses are:

- The kinetics of antibody production is accelerated in secondary responses as compared to primary responses.

- The immunoglobulin class of synthesised antibodies is different, with IgM initially and predominantly produced in primary responses, and IgG as the antibodies found in secondary responses.

- The magnitude of secondary responses is greater than that of primary responses, with much higher antibody titres detected in secondary responses.

- The affinity of antibodies is greater in secondary than in primary responses.

Despite these critical differences, only the primary antibody response is usually recommended to be explored in non-clinical immunotoxicity evaluation, even though it could seem more logical from the perspective of immunotoxicity risk assessment, to explore the secondary rather than the primary antibody response, as antibody responses mounted by human beings, except small children, are essentially secondary responses.

Very few studies were conducted to compare the influence of a given drug treatment or chemical exposure on the primary and secondary antibody responses in the same experimental conditions. When available, comparative results suggest that the secondary response is less sensitive than the primary response to chemical exposure. Therefore, current uncertainties on the practical relevance of the functional reserve capacity of the immune system could presumably be reduced by using more appropriate endpoints, such as the secondary antibody response.

The antigen

Antibody responses require B cells to mature into plasma cells to synthesise and release antibodies. Most, but not all, antibody responses are under the control ('help') of T lymphocytes. Therefore, two types of antigens, namely T-dependent and T-independent antigens, can be used to assess humoral immunity depending on whether T lymphocyte help is required to mount the antibody response. From the perspective of immunotoxicity risk assessment, antibody responses to T-dependent antigens are more relevant, as T-dependent antigens are by far the most common antigens. However, antibody responses to T-independent antigens can be helpful in mechanistic studies to identify immunotoxicants which specifically interfere with B lymphocyte function (as T lymphocytes have limited, if any role in the antibody responses against T-independent antigens).

A variety of T-dependent antigens are commonly used, including sheep red blood cells, tetanus toxoid, bovine serum albumin, ovalbumin, human gammaglobulin, and keyhole limpet haemocyanin (KLH). Sheep red blood cells have been largely used by immunologists, but KLH, a strong protein antigen, is also often recommended. Actually, the selection of the antigen seldom relies on objective data. The Fischer 344 rat was suggested to respond less strongly to antigens on sheep red blood cells, which could explain why some authors recommended KLH. Ovalbumin can induce IgM and IgG as well as IgE antibody response, depending on the experimental protocol used. In addition, ovalbumin as well as sheep erythrocytes and KLH, depending on the dose and route of administration, can induce either humoral or cellular immune response.

T-independent antigens, although seldom included in non-clinical immunotoxicity evaluation, include DNP-Ficoll, TNP-*Escherichia coli* lipopolysacharide (LPS), polyvinyl-pyrrolidone and flagellin.

Measurement of specific antibody levels

Typically, antibody titres/levels are measured 7–10 days after the injection of the antigen to rodents. The humoral response to an antigen can be measured from the circulating titres/levels of specific antibodies directed against this antigen. Until recently, antibody titres were generally measured using haemagglutination, complement lysis or antibody precipitation. Immunoassays, such as ELISAs, were considered to be more suitable by some investigators (Vos *et al.*, 1979); this however remains to be convincingly established, and ELISAs for the assay of specific antibody levels have not yet gained wide acceptance in immunotoxicity evaluation in contrast to the plaque-forming cell assay. However, the status of ELISAs could evolve due to current efforts paid to improving available techniques and validating results.

ELISA

ELISA (enzyme-linked immunoadsorbent assay) techniques are useful to assay antibodies or antigens whether the antibody or the antigen is bound to a solid phase (Law, 1996). The binding of the antibody with the antigen can be detected by an enzymatic reaction. Among the various variations of ELISA techniques, two are more interesting. The direct methods use an antigen bound non-covalently to a solid phase after the addition of specific serum and revealtion by an enzyme-labelled antibody. After a second washing, the substrate is added and the reaction between the bound enzyme and the substrate induces a coloured reaction that can be detected by spectrophotometry, the results being compared to a standard curve. This is a very quick, easy and sensitive method.

The competition technique is the mixing of a soluble antigen and a limited amount of antibody. The mixture is placed in wells covered by the antigen. The amount of antibody covered with the antigen during the second step, is proportional to the amount of complexes formed during the first step and hence the concentration of serum antibody.

Although ELISA techniques are very widely used in clinical immunology, their use in non-clinical immunotoxicology evaluation is not widespread for investigation of specific humoral immunity. The lack of good-quality reagents easily available on the market is undoubtedly a major limitation to the use of this technique nowadays, but much progress has been achieved recently (Burns *et al.*, 1996; Exon and Talcott, 1995; Heyman *et al.*, 1984; Van Loveren *et al.*, 1991).

ELISPOT

The enzyme-linked immunospot assay (ELISPOT) is sometimes considered an attractive alternative to the plaque-forming cell assay.

This test includes three steps (Czerkinsky *et al.*, 1983; Kawabat, 1995): the binding of the antigen to a solid phase; the incubation of antibody-producing cells; and the detection of antigen–antibody in antibody-producing cells by a coloured reaction (e.g. alkaline phosphatase = blue colour, or horseradish peroxidase = red colour). The number of antigen-producing cells is counted with a small-magnification microscope (10×–30×). This technique could have a better sensitivity and specificity than the plaque-forming cell assay. Moreover, the numeration of both IgM and IgG antibody-producing cells can be performed simultaneously, whereas the indirect plaque-forming cell assay (to measure IgG antibody-producing cells) is very difficult to perform in rats.

Other techniques

Several additional techniques are far less commonly used, either because they are considered obsolete today, or because they are more difficult, time-consuming or require expensive equipment and/or skilled technical staff.

Haemagglutination

Immunological agglutination is the hallmark of the formation of particle clusters suspended in a saline medium after an antigen–antibody reaction. The term haemagglutination is appropriate only when red blood cells are used as the antigen. Direct or active haemagglutination is related to the presence of agglutinating antibodies to components of the external surface of red cells. The antibodies are essentially IgM, as IgG are 10 to 100 times less agglutinating than IgM. The intravenous or intraperitoneal injection of washed sheep red blood cells to mice or rats induces a primary antibody response, which can be measured after 7–10 days (Dietrich, 1966). Serial serum dilutions are incubated with sheep red blood cells. When agglutinating antibodies (haemagglutins) are present, a pellet is formed at the bottom of an haemolysis tube or microwell plate. The higher titre associated with readable agglutination is determined. Results are usually expressed as $\log_2$ titre. Passive haemagglutination refers to the use of red blood cells previously coupled to protein (using tannic acid) or other small molecules (such as glutaraldehyde or chromic chloride).

This technique is largely considered obsolete despite several advantages. No sophisticated knowhow or expensive equipment are required. Samples can be kept at −15°C for a long period of time. The micromethod requires only minute amounts of sera (several microlitres). However, major limitations are the poor sensitivity and reproducibility, so that the observed titre could vary from one to two simply due to reading errors.

Radioimmunoassays

Radioimmunological techniques have a much greater sensitivity (the detection threshhold is approximately 0.01 µg/ml). They use intrinsically radiolabelled antigens. A standard serum of known potency is assayed in parallel to calibrate the antibody level. The major drawback is related to the technical difficulties to implement these techniques: experienced staff, expensive equipment, use of radioactive isotopes, etc. Therefore, radioimmunoassays are generally not considered for use in non-clinical immunotoxicity evaluation, except when antibodies directed againt the test article are suspected to be raised.

The plaque-forming cell assay

The enumeration of plaque-forming cells is the most frequently used technique to explore an immunosuppressive effect on humoral immunity (Holsapple, 1995). The major value

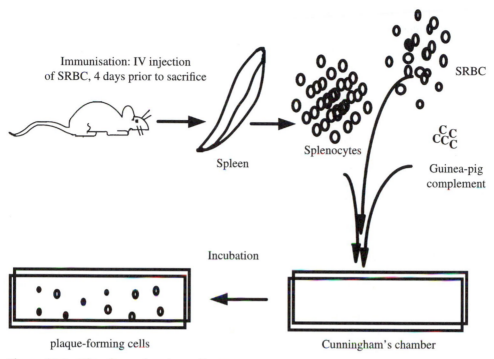

Figure 10.1 The plaque-forming cell assay

of this technique is to assess the antibody response (B lymphocytes), which is under the control of T lymphocytes, after antigen presentation by macrophages. The plaque-forming cell assay investigates the production of specific antibodies by antibody-producing cells following a primary or secondary immunisation. Direct (IgM) or indirect (IgG) plaque-forming cells can be measured (Jerne *et al.*, 1974). The most frequently used antigen is sheep red blood cells, but other antigens, such as KLH or tetanus toxoid, can be used after previous coupling using tannic acid, chromic choride or carbodiimide.

The technique initially described by Jerne and Nordin (1963) was later simplified by Cunningham (1965). The antigenic suspension is injected by the intravenous or intra-peritoneal route to rodents. The animals are killed after 4–5 days (direct plaque assay) or 7–10 days (indirect plaque assay) in order to take into account the kinetics of IgM and IgG production, respectively. The spleen is removed and a splenocyte suspension is incubated in agarose (Jerne's technique) or over a slide (Cunningham's technique) with sheep red blood cells and guinea-pig serum. At the end of the incubation period, clear areas are seen due to red blood cell lysis around spleen cells (haemolytic plaques). The number of plaques can be counted with the naked eye or at a very small magnification. The number of plaques is usually given for 10^5 or 10^6 splenocytes. When anti-IgG serum is added during the incubation period (indirect plaque assay), it is possible to measure the IgG response. However, indirect plaque assay is extremely difficult to perform in the rat. A technique combining ELISA and plaque assay has been proposed (Sedgwick and Holt, 1986).

B lymphocyte proliferation assay

B lymphocytes, as with all lymphocytes, have the property to proliferate when cultured. B lymphocyte proliferation is performed according to the same experimental conditions as T lymphocyte proliferation (see Chapter 11). *Escherichia coli* lipopolysaccharide (LPS) is the most commonly used B lymphocyte mitogen. However, *Salmonella typhi* LPS is the preferred B cell mitogen in rats (Smialowicz *et al.*, 1991).

References

BURNS, L.A., DUWE, R.L., JOVANOVIC, M.L., SEATON, T.D., JEAN, P.A., GALLAVAN, R.H., *et al.* (1996) Development and validation of the antibody-forming cell response as an immunotoxicological endpoint in the guinea-pig. *Toxicol. Meth.*, **6**, 193–212.

CUNNINGHAM, A.J. (1965) A method of increased sensitivity for detecting single antibody-forming cells. *Nature*, **207**, 1106–1107.

CZERKINSKY, C.C., NILSSON, L.A., NYGREN, H., OUTCHERLONY, O. and TARKOWSKI, A. (1983) A solid-phase enzyme-linked immunospot (ELISPOT) assay for enumeration of specific antibody-secreting cells. *J. Immunol. Meth.*, **65**, 109–121.

DIETRICH, F.M. (1966) Inhibition of antibody formation to sheep erythrocytes by various tumour-inhibiting chemicals. *Int. Arch. Allergy*, **29**, 313–328.

EXON, J.H. and TALCOTT, P.A. (1995) Enzyme-linked immunosorbent assay (ELISA) for detection of specific IgG antibody in rats. In: *Methods in Immunotoxicology, vol. 1* (Burleson, G., Dean, J.H. and Munson, A.E., eds), pp. 109–124. New York: Wiley-Liss.

HEYMAN, B., HOLMQUIST, G., BORWELL, P. and HEYMAN, U. (1984) An erythrocyte-linked immunosorbent assay for measuring anti-sheep erythrocyte antibodies. *J. Immunol. Meth.*, **68**, 193–204.

HOLSAPPLE, M.P. (1995) The plaque-forming cell (PFC) response in immunotoxicology: an approach to monitoring the primary effector function of B lymphocytes. In: *Methods in Immunotoxicology, vol. 1* (Burleson, G., Dean, J.H. and Munson, A.E., eds), pp. 71–108. New York: Wiley-Liss.

JERNE, N.K. and NORDIN, A.A. (1963) Plaque formation in agar by single antibody producing cells. *Science*, **140**, 405.

JERNE, N.K., HENRY, C., NORDIN, A.A., FUJI, H., KOROS, A.M.C. and LEFKOVITS, I. (1974) Plaque forming cells: methodology and theory. *Transplant. Rev.*, **18**, 130–191.

KAWABAT, T.T. (1995) Enumeration of antigen-specific antibody-forming cells by the enzyme-linked immunospot (ELISPOT) assay. In: *Methods in Immunotoxicology, vol. 1* (Burleson, G., Dean, J.H. and Munson, A.E., eds), pp. 125–135. New York: Wiley-Liss.

LADICS, G.S., SMITH, C., HEAPS, K., ELLIOTT, G.S., SLONE, T.W. and LOVELESS, S.E. (1995) Possible incorporation of an immunotoxicological functional assay for assessing humoral immunity for hazard identification purposes in rats on standard toxicology study. *Toxicology*, **96**, 225–238.

LAW, B. (1996) *Immunoassay. A Practical Guide.* London: Taylor & Francis.

SEDGWICK, J.D. and HOLT, P.G. (1986) ELISA plaque assay for the detection and enumeration of antibody-secreting cells. An overview. *J. Immunol. Meth.*, **87**, 37–44.

SMIALOWICZ, R.J., RIDDLE, M.M., LUEBKE, R.W., COPELAND, C.B., ANDREWS, D., ROGERS, R.R., *et al.* (1991) Immunotoxicity of 2-methoxyethanol following oral administration in Fischer 344 rats. *Toxicol. Appl. Pharmacol.*, **109**, 494–506.

TEMPLE, L., KAWABATA, T.T., MUNSON, A.E. and WHITE, K.L. (1993) Comparison of ELISA and plaque-forming cell assays for measuring the humoral response to SRBC in rats and mice treated with benzo[a]pyrene or cyclophosphamide. *Fund. Appl. Toxicol.*, **21**, 412–419.

VAN LOVEREN, H., VERLAAN, A.P.J. and VOS, J.G. (1991) An enzyme-linked immunosorbent assay of anti-sheep red blood cell antibodies of the classes M, G and A in the rat. *Int. J. Immunopharmac.*, **13**, 689–695.

VOS, J.G., BUYS, J., HANSTEDE, J.G. and HAGENAARS, A.M. (1979) Comparison of enzyme-linked immunosorbent assay and passive haemagglutination method for quantification of antibodies to lipopolysaccharide and tetanus toxoid in rats. *Infect. Immun.*, **24**, 798–803.

11

Assays of Cell-Mediated Immunity

Cell-mediated immunity is the second arm of specific immune responses. Cell-mediated immunity is evoked by T lymphocytes, but macrophages and to a lesser extent polymorphonuclear leukocytes also play a role in these reactions. The main manifestations of cell-mediated immunity are delayed type hypersensitivity, organ transplant rejection, tumour immunity, and resistance to a variety of infectious pathogens, such as parasites and intracellular bacteria. The most recent trend in immunological research gave much emphasis on this aspect of immune responsiveness. Available models to explore cell-mediated immunity include *in vivo* tests, such as delayed-type hypersensitivity, and to a much lesser extent in the context of immunotoxicity evaluation, allograft rejection and graft-versus-host reaction, and *in vitro* or *ex-vivo* tests, such as lymphocyte proliferative responses and more recently cytokine production (Luster *et al.*, 1982).

In vivo assays of cell-mediated immunity

Although they have often been neglected for the benefit of *in vitro* tests, *in vivo* models of cell-mediated immunity have the advantage of integrating the immune response at the level of the whole animal, and therefore can take into account indirect influences on the immune response, such as those derived from neurological or endocrine adverse effects of xenobiotics. When compared to *in vitro* tests however, *in vivo* models proved as predictive (Luster *et al.*, 1992).

Delayed-type hypersensitivity

Delayed-type hypersensitivity requires the specific recognition of a given antigen by activated T lymphocytes, which subsequently proliferate and release cytokines; these in turn increase vascular permeability, induce vasodilatation and macrophage accumulation, and finally antigen destruction. Delayed-type hypersensitivity reactions are good correlates of cell-mediated immunity, even though some authors claimed these models are less sensitive than *in vitro* methods. Available test models include classical delayed-type hypersensitivity and contact sensitivity models.

Classical delayed-type hypersensitivity

This is the most commonly used *in vivo* model for assessment of cell-mediated immunity (Henningsen *et al.*, 1984). Overall, delayed-type reactions have the advantage of exploring cell-mediated immunity in a living animal, but they are often considered to be less sensitive and poorly reproducible, if not in the hands of experienced staff. To mount experimentally a delayed-type reaction, three distinctive phases are absolutely required: the sensitising phase which corresponds to the single or repeated administration of the antigen; the rest period of varying duration; and the eliciting phase which corresponds to the readministration of the same antigen.

The antigen is generally injected by the subcutaneous or intradermal route to induce hypersensitivity. Sheep red blood cells are the most commonly used antigens (particularly in rodents), but any T-dependent antigens, such as keyhole limpet haemocyanin (KLH), toxoid anatoxin, *Listeria monocytogenes* or ovalbumin, can be used (Holsapple *et al.*, 1984; Lagrange *et al.*, 1974; Vos *et al.*, 1980). In many instances, the route and dose of the sensitising administration have been largely determined empirically, which explains the variety of experimental protocols which can be found in the literature. The duration of the rest period is also quite variable, but most often between 7 and 14 days, although a longer duration, e.g. 21 days, is sometimes used. Within 24 to 48 hours after the eliciting injection, the delayed-type hypersensitivity reaction is observed with local inflammation, redness, induration and/or oedema as the most typical signs, depending on the site of the eliciting injection.

In most instances, the reaction is measured as the increased thickness in hindfoot pad of a rodent, which reflects the magnitude of the delayed-type hypersensitivity response (Van Loveren *et al.*, 1984). Using a dial caliper, the foodpad thickness is measured immediately prior to the eliciting injection into the hind footpad, then after 24 and/or 48 hours, sometimes later (the reaction normally peaks at 48 hours). This technique can easily be performed in mice and rats. In an attempt to improve the accuracy and reproducibility of this technique, various improvements have been proposed (Vadas *et al.*, 1975), in particular methods using ^{125}I iodine diffusion to measure increased capillary permeability and oedema due to the delayed-type hypersensitivity reaction, or tritiated thymidine incorporation. In fact, these time-consuming and complex alternatives failed to prove their superiority to previous methods.

Less often, the delayed-type hypersensitivity reaction is assessed using intradermal readministration of the antigen to induce a cutaneous reaction similar to that seen in human skin testing, namely erythema with or without oedema. However, this model, which was used in the guinea-pig, is not sensitive and reproducible enough and should not be recommended. Recently, attempts have been made to develop this technique in monkeys (Bleavins and De La Iglesia, 1995).

Contact sensitivity

Contact hypersensitivity is a variety of delayed-type hypersensitivity, in which the antigen, essentially a hapten, is taken and processed by Langerhans cells, then presented to T CD4$^+$ lymphocytes. Potent contact sensitisers, such as picryl chloride, oxazolone and dinitrofluorobenzene (DNFB) in the mouse, and dinitrochlorobenzene (DNCB) and picryl chloride in the guinea-pig are used to induce a contact sensitivity reaction, the intensity of which can be shown to be modulated (decreased or augmented) by drug treatment or chemical exposure of the tested animals (Descotes and Evreux, 1982; Descotes *et al.*, 1985).

Because potent contact sensitisers are used, the experimental protocol is simplified compared to conventional sensitisation assays designed to assess the sensitising potential of chemicals (see Chapter 15). Sensitisation is performed topically on the shaved abdomen or interscapular area. The elicitation is achieved by topical application on the shaved abdomen (guinea-pig, mouse) or the ear (mice) of a concentration which had previously been shown to induce no primary irritation. In guinea-pigs, the magnitude of the response is measured semi-quantitatively based on the intensity of the erythema and the presence or lack of oedema. In mice, ear thickness is measured immediately before elicitation and 24–48 hours later using an engineer micrometer or more appropriately a dial caliper. The increase in ear thickness is a good indicator of delayed-type hypersensitivity. A radioisotopic assay was also proposed (Lefford, 1974).

Contact hypersensitivity models, although less often used in immunotoxicity evaluation give results similar to those obtained either with 'typical' delayed-type hypersensitivity or contact skin reactions. However, one major limitation is the limited sensitivity of rats to contact sensitisers.

Other in vivo assays of cell-mediated immunity

Delayed-type hypersensitivity, either typical or contact, reactions are by far the most commonly used *in vivo* assays to study the influence of immunotoxicants on cell-mediated immunity. Other assays can nevertheless be considered in particular situations.

Allograft rejection

Allograft models are no longer much used for non-clinical immunotoxicity evaluation. Skin grafts have exceptionally been used in this context as they are more adapted to immunopharmacological investigation.

Graft-versus-host disease

Graft-versus-host (GvH) disease results from the injection of histo-incompatible immune cells to an immature, immunocompromised or tolerant host: the donor's cells react with the host's antigens, whereas the host is unable to mount an immune response against the donor's cells (Krzystyniak *et al.*, 1992). This phenomenon is essentially cell-mediated.

To induce GvH disease experimentally, immune cells from a syngenic donor (mouse or rat) are injected to a newborn or irradiated host. F1 hybrid hosts can also be used. Depending on the modalities of donor cells injected, the GvH disease can be either acute or chronic, systemic or local. However, these models are essentially used for the evaluation of new immunosuppressive agents.

In vitro assays of cell-mediated immunity

In vitro (or *ex-vivo*) assays used to be preferred by immunologists to explore cell-mediated immunity. It is therefore not surprising that such assays have been increasingly used in non-clinical immunotoxicity evaluation.

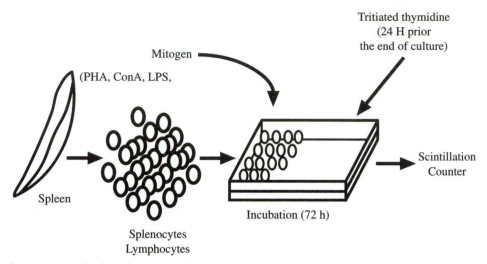

Figure 11.1   The lymphocyte proliferative assay

Lymphocyte proliferative response assays

These assays take advantage of the capacity of cultured lymphocytes to proliferate (Smialo-wicz, 1995). *In vitro* proliferation is a well-recognised property of T lymphocytes and it has been shown to be a good correlate of cell-mediated immunity.

At the end of the drug treatment or chemical exposure, animals are killed to collect spleen lymphocytes. It is also possible to collect lymphocytes from peripheral blood, which to a certain extent, avoids the killing of animals, even though the large volume of blood that is usually required is not compatible with the animal's survival, especially when rodents are used. Lymphocytes can also be collected from lymph nodes. When lymphocytes are collected from treated animals, this is an *ex-vivo* assay; when increasing concentrations of the test compound are added to cultured lymphocytes collected from untreated animals, this is an *in vitro* assay.

Lymphocytes are cultured for a varying period of time (often 72 hours). Four to 24 hours prior to the end of the culture, tritiated thymidine is usually incorporated. When lymphocytes proliferate, they incorporate tritiated thymidine into DNA, so that the amount of incorporated radioactivity in cultured lymphocytes is a correlate of lymphocyte pro-liferation and hence of cell-mediated immunity. Other methods have been proposed to quantify lymphocyte proliferation. These are essentially colorimetric methods using rea-gents, such as MTT (Mosmann, 1983), XTT (Roehm *et al.*, 1991) and bromodeoxyuridine (BrdU). However, the proliferation indices obtained with colorimetric methods are usu-ally much smaller than when tritiated thymidine is used.

Although lymphocytes can proliferate when cultured, proliferation has to be enhanced to achieve measurable levels. Lectins (mitogens) have the capacity to trigger lymphocyte proliferation non-specifically through interactions with cell surface binding sites. The most commonly used mitogens are phytohaemagglutinin (PHA) and chiefly concanavalin A (ConA), which cause T lymphocytes to proliferate; lipopolysaccharides to induce B lymphocyte proliferation; and pokeweed mitogen (PWM), which induces the prolifera-tion of both B and T lymphocytes. Lymphocyte proliferation can also be induced when using allo-antigens. The animals are sensitised *in vivo* by an antigen, such as tuberculin, and lymphocyte proliferation is induced *in vitro* by the addition of the *Bacillus tuberculis* extract PPD (protein purified derivative).

The proliferative response of T lymphocytes can also be produced by surface antigens on allogeneic cells. The mixed lymphocyte response assay therefore measures responses which are involved in graft rejection and graft-versus-host reactions. In this assay, 'responder' cell suspensions from the single spleen of exposed animals are prepared and incubated in the presence of 'stimulator' pooled splenocytes from control histo-incompatible animals, after inactivation by treatment with mitomycin C. Eighteen hours before termination of the culture, tritiated thymidine is added for incorporation into proliferating lymphocytes.

Mitogen-induced lymphocyte proliferative assays are less often utilised than they used to be in the past, essentially because of their lack of reproducibility and sensitivity. The mixed lymphocyte reaction gained some favour recently, but it is not certain whether results obtained with this technique are more reliable and reproducible.

T lymphocyte cytotoxicity assay

Cytotoxic T lymphocytes (CTL) are T-CD8$^+$ lymphocytes specifically cytotoxic to target cells. Although cytotoxic T lymphocytes can directly destroy their targets, CTL lysis is a major histocompatibility complex (MHC)-restricted process requiring prior sensitisation for T lymphocytes to proliferate and differentiate into effector cytotoxic cells via the sequential influence of several cytokines. The *in vitro* induction of CTL from the splenocytes of treated animals is a widely used model to evaluate cell-mediated immunity (House and Thomas, 1995). Splenocytes of treated and control animals are co-cultured usually for five days with target cells. At the end of the culture, the splenocytes are washed and added to fresh radio-labelled target cells. Cytotoxicity is measured from the amount of radioactivity (the radiolabel is usually ^{51}Cr) into the supernatant after a four-hour incubation period. Commonly used target cells are P815 mastocytoma cells in mice and Fu-G1 tumour cells in rats.

Cytokine assays

In the past decade, a number of assays have been described to measure cytokines in body fluids or tissues (Bienvenu *et al.*, 1998; House, 1995). Very low concentrations of cytokines (<10 pg/ml) are associated with pleiotropic effects on numerous target cells. They can be found under multiple molecular forms, such as monomers/polymers, glycosylated derivatives, precursors, and degradation products, with different assay behaviours. In addition, a number of cytokine inhibitors have been described. As the half-life of cytokines is very short (<10 minutes in most instances), the conditions of sampling and storage of biological fluids for cytokine measurements are very important.

Bioassays or immunoassays can be used to measure cytokines. In bioassays, the biological activity of a cytokine is tested on target cells in comparison with a standard cytokine preparation to establish a dose–response curve. Bioassays are based on various biological responses, such as proliferation, cytotoxicity, or antiviral activity (Thorpe *et al.*, 1992). The major drawbacks of bioassays are the lack of specificity and the poor reproducibility. Furthermore, they are time-consuming. Immunoassays replaced bioassays in clinical practice because of their high specificity. Radio- and enzymo-immunoassays (ELISA) use a first (monoclonal) antibody for the capture of the cytokine, and a second (monoclonal or polyclonal) labeled antibody. Many commercial kits are available with a detection limit usually around 5 pg/ml.

The capacity of cells to produce cytokines can be explored by ELISPOT or whole blood models. In ELISPOT, a cell suspension is placed on a Petri dish coated with antibodies directed against the cytokine of interest (Czerkinsky *et al.*, 1991). After incubation and washing, cytokine-producing cells are captured by the antibody, and the fixation of the cells is visualised by a second anticytokine antibody labelled by an enzyme. In whole blood models, the mitogens lipopolysaccharide and phytohaemagglutinin are generally used as stimuli for cytokine synthesis, particularly pro-inflammatory cytokines.

The tissue distribution, number and nature of cytokine-producing cells can be investigated through the detection of either cellular cytokines or cytokine mRNA by *in situ* hybridisation or reverse transcriptase polymerase chain reaction (RT-PCR). Immunohistochemistry allows cytokine detection within tissues with the phenotyping of secreting cells. The major steps are fixation, permeabilisation of cells, and addition of specific anticytokine antibodies. The selection of antibodies is crucial. The intracellular detection of cytokines is possible on peripheral blood cells, bronchoalveolar lavage, cytopuncture or tissue biopsies. The binding of antibodies to cytokines is visualised by immuno-fluorescent or immuno-enzymatic techniques. Double or triple-labelling can be used to detect cell surface antigens and identify the phenotype of cytokine-secreting cells. The main limitation is the low concentration of intracellular cytokines. The detection of intracellular cytokines by flow cytometry on cell suspensions is possible by using fluorescent anticytokine antibodies. The major advantage is to characterise the phenotype and intracytoplasmic cytokine content of the cells in the same experiment.

The detection of cytokine mRNA on tissue fragments or cell suspensions is complementary to immunohistochemistry. The best results are obtained with riboprobes (single-strand RNA probes) or oligonucleotides. Competitive RT-PCR is one of the most interesting tools. Results are expressed as cytokine mRNA copies per gram of total cellular RNA. By using a multispecific standard, it is possible to study the messengers for several cytokines in the same experiment.

The discovery that cytokines play pivotal roles in many aspects of the immune response, is causing profound changes in the functional exploration of lymphocytes. These advances have reached the area of immunotoxicology. However, data are still relatively scarce and largely limited to research.

References

BIENVENU, J., MONNERET, G., GUTOWSKI, M.CL. and FABIEN, N. (1998) Cytokine assays in human sera and tissues. *Toxicology*, **129**, 55–61.

BLEAVINS, M.R. and DE LA IGLESIA, F.A. (1995) Cynomolgus monkeys (*Macaca fascicularis*) in preclinical immune function safety testing: development of a delayed-type hypersensitivity procedure. *Toxicology*, **95**, 103–112.

CZERKINSKY, C., ANDERSSON, J., FERRUA, B., NORSTROM, I., QUIDING, M., ERIKSSON, K., *et al.* (1991) Detection of human cytokine-secreting cells in distinct anatomical compartments. *Immunol. Rev.*, **119**, 5–22.

DESCOTES, J. and EVREUX, J.CL. (1982) Effect of chlorpromazine on contact hypersensitivity to DNCB in the guinea-pig. *J. Neuroimmunol.*, **2**, 21–25.

DESCOTES, J., TEDONE, R. and EVREUX, J.CL. (1985) Immunotoxicity screening of drugs and chemicals: value of contact hypersensitivity to picryl chloride in the mouse. *Meth. Find. Exp. Clin. Pharmacol.*, **7**, 303–305.

HENNINGSEN, G.M., KOLLER, L.D., EXON, J.H., TALCOTT, P.A. and OSBORNE, C.A. (1984) A sensitive delayed-type hypersensitivity model in the rat for assessing in vivo cell-mediated immunity. *J. Immunol. Meth.*, **70**, 153–165.

HOLSAPPLE, M.P., PAGE, D.G., SHOPP, G.M. and BICK, P.H. (1984) Characterization of the delayed hypersensitivity response to a protein antigen in the mouse. *Int. J. Immunopharmac.*, **6**, 399–405.

HOUSE, R.V. (1995) Cytokine bioassays for assessment of immunomodulation: applications, procedures, and practical considerations. In: *Methods in Immunotoxicology, vol. 1* (Burleson, G., Dean, J.H. and Munson, A.E., eds), pp. 251–276. New York: Wiley-Liss.

HOUSE, R.V. and THOMAS, P.T. (1995) *In vitro* induction of cytotoxic T lymphocytes. In: *Methods in Immunotoxicology, vol. 1* (Burleson, G., Dean, J.H. and Munson, A.E., eds), pp. 159–171. New York: Wiley-Liss.

KRZYSTYNIAK, K., PANAYE, G., DESCOTES, J. and REVILLARD, J.P. (1992) Changes in lymphocyte subsets during acute local graft-versus-host reaction in H-2 incompatible murine F1 hybrids. *Immunopharmacol. Immunotoxicol.*, **14**, 865–882.

LAGRANGE, P., MACKANESS, G.B. and MILLER, T.E. (1974) Influence of dose and route of antigen injection on the immunological induction of T cells. *J. Exp. Med.*, **139**, 528–542.

LEFFORD, M.J. (1974) The measurement of tuberculin hypersensitivity in rats. *Int. Arch. Allergy Clin. Immunol.*, **47**, 570–585.

LUSTER, M.I., DEAN, J.H. and BOORMAN, G.A. (1982) Cell-mediated immunity and its application to toxicology. *Environ. Health Perspect.*, **43**, 31–36.

LUSTER, M.I., PORTIER, C., PAIT, D.G., WHITE, K.L., GENNINGS, C., MUNSON, A.E. and ROSENTHAL, G.J. (1992) Risk assessment in immunotoxicology. I. Sensitivity and predictability of immune tests. *Fund. Appl. Toxicol.*, **18**, 200–210.

MOSMANN, T. (1983) Rapid colorimetric assay for cellular growth and survival: application to proliferation and cytotoxicity assays. *J. Immunol. Meth.*, **65**, 55–63.

ROEHM, N.W., RODGERS, G.H., HATFIELD, S.M. and GLASEBROOK, A.L. (1991) An improved colorimetric assay for cell proliferation and viability using the tetrazolium salt XTT. *J. Immunol. Meth.*, **142**, 257–265.

SMIALOWICZ, R.J. (1995) *In vitro* lymphocyte proliferation assays: the mitogen-stimulated response and the mixed-lymphocyte reaction in immunotoxicity testing. In: *Methods in Immunotoxicology, vol. 1* (Burleson, G., Dean, J.H. and Munson, A.E., eds), pp. 197–210. New York: Wiley-Liss.

THORPE, R., WADHWA, M., BIRD, C.R., MIRE-SLUIS, A.R. (1992) Detection and measurement of cytokines. *Blood Reviews*, **6**, 133–148.

VADAS, M.A., MILLER, J.F.A., GAMBLE, J. and WHITELAW, A. (1975) A radioisotopic method to measure delayed type hypersensitivity in the mouse. *Int. Arch. Allergy Appl. Immunol.*, **49**, 670–692.

VAN LOVEREN, H., KATO, K., RATZLAFF, R.E., MEADE, R., PTAK, W. and ASKENASE, P.A. (1984) Use of micrometers and calipers to measure various components of delayed-type hypersensitivity ear swelling reactions in mice. *J. Immunol. Meth.*, **67**, 311–319.

VOS, J.G., BOERKAMP, J., BUYS, J. and STEERENBERG, P.A. (1980) Ovalbumin immunity in the rat: simultaneous testing of IgM, IgG and IgE response measured by ELISA and delayed-type hypersensitivity. *Scand. J. Immunol.*, **12**, 289–295.

12

Assays of Non-Specific Defences

In addition to antigen-specific defences, non-specific defence mechanisms play a major role in the protection of the host's integrity, particularly against microbial pathogens. A number of non-specific mechanisms, either humoral or cellular, have been suggested or demonstrated to be involved. Among humoral mechanisms, serum enzymes and proteins, such as C-reactive protein, lyzosyme, caeruloplasmin, alpha-2-macroglobulin, and alpha-1-antitrypsin, have been subject to extensive investigation in the past, but their predictive value has never been carefully assessed and they are no longer recommended or included in routine non-clinical immunotoxicity evaluation.

Therefore, cellular mechanisms of non-specific defences are of primary interest. Neutrophils and mononuclear phagocytes, namely monocytes and macrophages, play a major role in non-specific immune responses and inflammatory reactions. In addition, much emphasis has been placed on natural killer cell activity in non-clinical immunotoxicity evaluation.

Assays of phagocytosis

General considerations

Phagocytosis

Five major steps can be identified in the process leading to the destruction of microorganisms by phagocytes. The generic term 'phagocytosis' is often used to characterise the whole process, although strictly speaking, phagocytosis should only refer to the engulfment or ingestion step.

First, phagocytes have to move in order to reach their target. The two critical components of phagocyte locomotion are the direction (chemotaxis) and speed (chemokinesis). A number of chemoattractants have been identified with both chemotactic and chemokinetic properties: the main natural substances are proteins and peptides released by bacteria (such as the chemoattractant N-formyl-methionyl-leucyl-phenylalanine or FMLP), by-products of the complement activation cascade (such as C5a), and various derivatives of activated inflammatory proteins (such as kallikrein and fibrinopeptide B). The locomotion

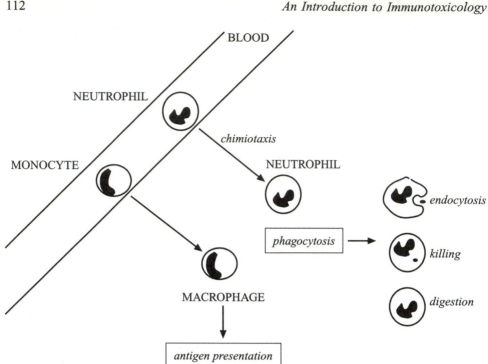

Figure 12.1 The main steps of phagocytosis

of phagocytes also depends on their adherence to the surface upon which they are moving. Adherence is increased by FMLP and adhesion proteins, such as LFA-1.

The second step of phagocytosis is the adhesion of phagocytes to their targets. A large number of membrane receptors have been identified. The best characterised and functionally most important receptors are Fcγ receptors I, II and III, which recognise the Fc domain of IgG, complement receptor 1, which preferentially binds C3b, and complement receptor 3 (C11b/18), and IgA as well as IgE (CD23) receptors.

The third step is the ingestion of microorganisms after adhesion to phagocytes. This is a very complex process involving marked conformational and functional changes in the phagocyte membrane and the cytoskeleton, which are still the subject of extensive investigation. Phagocytes are activated and this is evidenced by major intracellular metabolic changes. Several biochemical pathways have been identified to play a major role in the destruction of pathogens. They involve acidic hydrolases, lysozyme, myeloperoxidase, and oxygen reactive intermediates. The fourth and final step of phagocytosis is the digestion of pathogens by enzymes present in phagolysozomes.

Phagocyte function in non-clinical immunotoxicity evaluation

The role of phagocytosis is primarily the removal of microorganisms and foreign bodies, but also the elimination of dead or injured cells. Phagocyte defects are associated with varied pathological conditions in humans (White and Gallin, 1986).

Because phagocytosis is such a complex process, a fairly large number of *in vivo* and *in vitro* techniques have been designed and proposed to investigate the negative influence of immunotoxicants on phagocytosis. Even though alterations in phagocyte function is certainly a major issue (Gardner, 1984), this aspect of immunotoxicology has so far not been paid much attention. In addition, it is often unclear what clinical consequences the observed changes in selected phagocyte functions are likely to have.

Limitations of phagocyte function assays

Several major limitations can be found in current assays of phagocytosis and macrophage function. Most assays are only helpful for investigation of highly specific aspects of the whole phagocytic process. Due to the lack of global assays, with the exception of the very few *in vivo* assays, information to be gained is unlikely to be relevant from the perspective of immunotoxicity risk assessment, unless several assays are combined. In addition, because available assays have essentially been used in the context of fundamental research, they usually have not been properly standardised and validated.

This chapter is an attempt to overview those techniques which could be more helpful for non-clinical immunotoxicity evaluation.

In vivo *assays*

Phagocytosis can be investigated *in vivo* using clearance assays. Following the intravascular injection of colloidal carbon (Halpern *et al.*, 1953), or labelled triolein, which are taken up by phagocytes, the blood concentration can be measured repeatedly over a predetermined period of time (usually 30–90 minutes). These early techniques are not sensitive, and are therefore no longer used for non-clinical immunotoxicity evaluation, despite the theoretical advantage of investigating phagocyte function globally in the whole animal.

A second type of clearance assay is the *Listeria monocytogenes* clearance assay. At 24–48 hours after the intravenous injection of *Listeria monocytogenes* organisms, viable bacteria can be counted in the blood, liver and spleen of mice or rats (Bradley, 1995). The majority of injected *Listeria* are indeed quickly removed and destroyed by macrophages, so that the determination of *Listeria monocytogenes* clearance is a good indicator of macrophage function.

In vitro *assays*

The five successive steps of phagocytosis can be studied using *in vitro* (or *ex-vivo*) assays.

Chemotaxis

Various techniques have been proposed, and their respective merits and limitations discussed (Bignold, 1988; Wilkinson, 1982; Zigmond and Lauffenburger, 1986).

- *Boyden's chambers* (Boyden, 1962). This technique is based on the progress of leukocytes in suspension through a polycarbonate or nitrocellulosis filter covered with a chemoattracting substance. By using multi-well chambers, it is possible to perform a large number of measurements from treated and control animals simultaneously (*ex-vivo* technique), or from leukocyte suspensions incubated with increasing concentrations of the test substance (*in vitro* technique). After several hours of incubation with the chemoattractant (usually FMLP), the distance travelled by leukocytes is measured, which reflects the magnitude of the chemotactic response. To measure this distance, the most common method is the migration front technique, but image analysis can also be used.

- *The agarose technique* (Nelson *et al.*, 1975). Because this technique is easier and quicker to perform, it is more often used. Leukocyte migration is studied in wells of 2–3 mm in diameter dug in agarose. Wells contain either the leukocyte suspension, the chemoattractant (FMLP) or a control substance. It is thus possible to measure the distance travelled by leukocytes.

Adhesion

Mononuclear leukocytes can bind to, and subsequently phagocytose foreign particles through a number of membrane structures (receptors). However, this aspect of phagocytosis is very seldom considered during non-clinical immunotoxicity evaluation.

Ingestion

Phagocyte adhesion induces the ingestion of the target so that it is often difficult to differentiate both steps. Ingestion can be studied *in vitro* following the incubation of leukocytes with various particles, such as sheep red blood cells, opsonised zymosan or bacteria (*Escherichia coli, Staphylococcus aureus, Saccharomyces cerevisiae*), and sephadex beads (Neldon *et al.*, 1995).

The phagocyte suspension and particles are incubated at 37°C for 1–2 hours. Cells are subsequently coloured for microscopic examination. Cells which have ingested particles are counted. This technique has several limitations: it is time-consuming and therefore poorly suited to routine toxicological evaluation; and it is poorly reproducible and relatively inaccurate owing to the experimental conditions (the phagocytic index is calculated from the number of cells which ingested three, four or five particles, so that a different phagocytic index can be obtained depending on the selected experimental conditions).

Killing and metabolic activation of phagocytes

The killing of microbial pathogens is the result of the oxidative burst in phagocytes. Phagocyte activation results in the release of varied substances, such as acidic hydrolases, lysozyme, superoxide anion and reactive oxygen intermediates (Rodgers, 1995). The activation of neutrophils is usually more pronounced than that of mononuclear phagocytes.

The various proposed methods (Qureshi and Dietert, 1995) study different aspects of the oxidative metabolism of phagocytes, and their respective predictive value has not been compared from the perspective of non-clinical immunotoxicity evaluation. Whatever the measured endpoint, the metabolic activation of phagocytes can be induced by zymosan or phorbol myristate acetate (PMA), or the ingestion of latex beads, sheep red blood cells or bacteria (such as *Saccharomyces cerevisiae*).

- *Tetrazolium nitroblue reduction test* (Hisadome *et al.*, 1990). In this test, the reduction of tetrazolium nitroblue by the superoxide anion produced during the metabolic activation of phagocytes, is measured. The colour of reduced tetrazolium nitroblue is changed from pale yellowish to dark blue, and the intensity of the reaction can be measured using spectrophotometry or colorimetry.

- *Chemiluminescence* (Fromtling and Abruzzo, 1985; Verdier *et al.*, 1993). This method measures the production of light photons by activated phagocytes. The activation of phagocytes is associated with the production of light photons which can be measured with a luminometer provided that the reaction is amplified by the addition of luminol or lucigenin. This method is a relatively faithful tool to study the metabolic responses of phagocytes previously exposed to drug treatment or chemicals. Unfortunately, rat neutrophils are relatively less sensitive to the effects of stimulating agents than dog, monkey and human neutrophils.

- *Ferricytochrome C reduction test.* Superoxide anion has the capacity to reduce ferricytochrome C, which can easily be detected using photometry (Rodgers, 1995). Neutrophil activation is induced by the incubation with zymosan.

- *Flow cytometry* is increasingly used to measure phagocytosis, and particularly the metabolic activation of phagocytes following ingestion of beads or non-specific stimulation by PMA (Model *et al.*, 1997).

Natural killer cell activity

Natural killer cell activity is measured by incubating target cells (such as YAK cells) which have been previously labelled with [51]chromium (Djeu, 1995). Labelled target cells are washed and co-cultured for 4–18 hours with mononuclear leukocytes including NK cells. At the end of the co-culture, the supernatant is removed and the radioactivity counted. There is a relationship between the level of radioactivity in the supernatant and the amount of released [51]chromium; and between the amount of released [51]chromium and the functional activity of NK cells which kill target cells, and which in turn release [51]chromium.

NK cell activity can also be measured using flow cytometry (Chang *et al.*, 1993).

The interlaboratory validation study performed under the auspices of the US National Toxicology Program showed that NK cell activity is a reliable endpoint. Therefore, NK cell activity is often used in immunotoxicity evaluation.

References

BIGNOLD, L.P. (1988) Measurement of chemotaxis of polymorphonuclear leukocytes in vitro. *J. Immunol. Meth.*, **108**, 1–18.

BOYDEN, S. (1962) The chemotactic effect of mixtures of antibody and antigen on polymorphonuclear leucocytes. *J. Exp. Med.*, **115**, 453–466.

BRADLEY, S.G. (1995) *Listeria* host resistance model. In: *Methods in Immunotoxicology, vol. 2* (Burleson, G., Dean, J.H. and Munson, A.E., eds), pp.169–179. New York: Wiley-Liss.

CHANG, L., GUSEWITCH, G.A., CHRITTON, D.B.W., FOLZ, J.C., LEBECK, L.K. and NEHLSEN-CANARELLA, S.L. (1993) Rapid flow cytometry assay for the assessment of natural killer cell activity. *J. Immunol. Meth.*, **166**, 45–54.

DJEU, J.Y. (1995) Natural killer activity. In: *Methods in Immunotoxicology, vol. 1* (Burleson, G., Dean, J.H. and Munson, A.E., eds), pp. 437–449. New York: Wiley-Liss.

FROMTLING, R.A. and ABRUZZO, G.K. (1985) Chemiluminescence as a tool for the evaluation of antimicrobial agents: a review. *Meth. Find. Exptl. Clin. Pharmacol.*, **7**, 493–500.

GARDNER, D.E. (1984) Alterations in macrophage functions by environmental chemicals. *Environ. Health Perspect.*, **55**, 343–358.

HALPERN, B.N., BENACERRAF, B. and BIOZZI G. (1953) Quantitative study of the granulopectic activity of the reticulo-endothelial system. I: the effect of the ingredients present in India ink and of substances affecting blood clotting *in vivo* on the fate of carbon particles administered intravenously in rats, mice and rabbits. *Br. J. Exp. Pathol.*, **34**, 426–457.

HISADOME, M., FUKUDA, T. and TERASAWA, M. (1990) Effect of cysteine ethylester hydrochloride (Cystanin) on host defense mechanisms (V): potentiation of nitroblue tetrazolium reduction and chemiluminescence in macrophages or leukocytes of mice and rats. *Japan J. Pharmacol.*, **53**, 57–66.

MODEL, M.A., KUKURUGA, M.A. and TODD, R.F. (1997) A sensitive flow cytometric method for measuring the oxidative burst. *J. Immunol. Meth.*, **202**, 105–111.

NELDON, D.L., LANGE, R.W., ROSENTHAL, R.J., COMMENT, C.E. and BURLESON, G.E. (1995) Macrophage nonspecific phagocytosis assays. In: *Methods in Immunotoxicology, vol. 2* (Burleson, G., Dean, J.H. and Munson, A.E., eds), pp. 39–57. New York: Wiley-Liss.

NELSON, R.D., QUIE, P.G. and SIMMONS, R.L. (1975) Chemotaxis under agarose: a new and simple method for measuring chemotaxis and spontaneous migration of human polymorphonuclear leukocytes and monocytes. *J. Immunol.*, **115**, 1650–1656.

QURESHI, M.A. and DIETERT, R.R. (1995) Bacterial uptake and killing by macrophages. In: *Methods in Immunotoxicology, vol. 2* (Burleson, G., Dean, J.H. and Munson, A.E., eds), pp. 119–131. New York: Wiley-Liss.

RODGERS, K. (1995) Measurement of the respiratory burst of leukocytes for immunotoxicologic analysis. In: *Methods in Immunotoxicology, vol. 2* (Burleson, G., Dean, J.H. and Munson, A.E., eds), pp. 67–77. New York: Wiley-Liss.

VERDIER, F., CONDEVAUX, F., TEDONE, R., VIRAT, M. and DESCOTES, J. (1993) In vitro assessment of phagocytosis. Interspecies comparison of chemiluminescence response. *Toxicol. in vitro*, **7**, 317–320.

WHITE, C.J. and GALLIN, J.I. (1986) Phagocyte defects. *Clin. Immunol. Immunopathol.*, **40**, 50–61.

WILKINSON, P.C. (1982) The measurement of leucocyte chemotaxis. *J. Immunol. Meth.*, **51**, 133–148.

ZIGMOND, S.H. and LAUFFENBURGER, D.A. (1986) Assays of leukocyte chemotaxis. *Annu. Rev. Med.*, **37**, 149–155.

13

Host Resistance Assays

Initially, host resistance assays were developed for use in the context of immuno-suppression. Such assays include experimental infections and implanted tumours. They have been shown to be useful tools to investigate the global consequences of functional alterations of the immune response due to drug treatment or chemical exposure (Dean *et al.*, 1982; Thomas and Sherwood, 1996).

As discussed earlier in this volume, the immune system utilises a very complex inter-play of both non-specific and specific defence mechanisms to protect the body against microbial invaders and neoplasias. As these mechanisms are most often intricated and redundant, it is difficult to predict whether one or several functional immune changes are likely to alter the global resistance of the host, and if so, to what extent. Interestingly though, correlations have been evidenced between given changes of the immune function and susceptibility to experimental infections or tumours (Luster *et al.*, 1994; Selgrade *et al.*, 1992). The potential value of host resistance models is therefore primarily to serve as an aid to interpret the results of functional immune assays, and to determine whether a drug treatment or chemical exposure associated with immune functional changes, is likely to affect the global resistance of the host. Based on our current knowledge of the clinical consequences of immunosuppression, the finding that a drug treatment or chem-ical exposure is associated with impaired resistance to microbial infections and experi-mental tumours is indeed the best way to provide evidence that the drug treatment or chemical exposure under consideration is actually immunotoxic. Despite the value of host resistance models in immunotoxicity risk assessment, limitations should also be emphas-ised, particularly the need to use fairly large numbers of animals.

The design of host resistance assays came out prior to the elaboration of preclinical immunotoxicological evaluation strategies, and they in fact proved essential to demon-strate the need for this evaluation. Although host resistance assays are generally utilised in the context of immunosuppression, there is no reason to believe they cannot also be helpful for the global assessment of other immunotoxic effects, such as unexpected immunostimulation or immunomodulation (Herzyk *et al.*, 1997).

Experimental infection models

The selection of an experimental infection model for use in non-clinical immunotoxicity evaluation is based on various criteria: the risk for contaminating the toxicology laboratory staff as well as animal housing facilities; the sensitivity and pathogenicitiy of microbial strains; the reproducibility of results; and the type of immune alterations (namely T or B lymphocyte, or phagocyte dysfunction) which are suspected to result in increased susceptibility to infections.

General considerations

Selection of the infectious model

In the process of selecting experimental infectious models for non-clinical immunotoxicity evaluation, several points have to be carefully considered:

- The pathogenicity of the experimental infection should be as close to that seen in man as possible, and the defence mechanisms in the animal species used should be identical, or at least very similar, to those involved in man. The use of 'exotic' infectious models is not recommended as the interpretation of results for immunotoxicity risk assessment is likely to be difficult when the pathogenesis of the infectious disease is not well-characterised. This is probably the reason why the number of recommended experimental infectious models has steadily decreased in recent years.

- Immunological mechanisms involved in the host's defence against the experimental infection should be relevant with regard to the risk assessment of the immunotoxicant under scrutiny. The use of experimental infections involving essentially humoral mechanisms, such as *Streptococcus pneumoniae* infection, is irrelevant for assessment of the consequences of a compound shown to affect primarily the cell-mediated immunity. On the contrary, it would be similarly irrelevant to assess the global consequences of a compound shown to depress humoral immunity by using an infectious model involving cell-mediated immunity, such as *Listeria monocytogenes* infection. However, it should be kept in mind that mechanisms involved in the host's defence against microbial pathogens are often intricated and complex, so that selecting the appropriate infectious model may not be easy. Using several infectious models may therefore be necessary when assessing chemicals with complex effects on the immune response. However, the cost-effectiveness of this evaluation would be reduced accordingly.

- The amount of inoculated pathogens should be small enough to avoid exceeding the defence capacity of the animal's immune system. Theoretically, the appropriate dose of infectious agent must be determined to produce the desired lethality, for instance the LD_{50} (Laschi-Loquerie *et al.*, 1987), which requires that several groups of animals are used to determine the appropriate conditions of infectious challenge. It is however obvious that microbial virulence should be maintained at a relatively constant level, which can be achieved with rigid quality control and the proper selection of the microbial pathogen (Thomas and Sherwood, 1996).

- For the purpose of adequate risk assessment, experimental infectious models should be conducted in the same animal species as the species used during functional immune studies. Many infectious models have been designed in the mouse, whereas fewer models are available in the rat.

Evaluation criteria

The most commonly used criterion to evaluate the resistance to experimental infections is mortality, probably because this is the simplest criterion. However, mortality is unlikely to be the most relevant endpoint, as infections are fortunately rarely lethal in man. The current lack of extensive epidemiological studies correlating immunotoxic exposures with the extent of morbidity and mortality due to infectious diseases in human populations, is a major limitation to a sound and objective extrapolation from animals to man.

Lethality is possibly a relevant endpoint when the tested xenobiotic is potently immunosuppressive, but this endpoint is likely to be too drastic when assessing an immunodepressive or immunostimulating agent. In these latter cases, the time interval between inoculation and the onset of initial clinical symptoms, the duration of clinical symptoms, and the use of biological or indirect endpoints can prove more relevant, provided that a careful standardisation and validation of the selected endpoints are performed. Such criteria are however less easily assessed and standardised than lethality.

Limitations of infectious models

Although theoretically experimental infection models are the best way to assess the global impact of functional immune changes on the host's defences against microbial pathogens, three major limitations in the use of these models can be identified:

- The animal's death for toxicity evaluation is less and less accepted in many countries for ethical reasons in animal care. Whatever the importance for public health to predict whether drug treatments or chemicals exposures are likely to impair host resistance against microbial infections, inoculating laboratory animals to induce deaths is unacceptable by a growing fraction of the population in the Western world.

- Because it is often impossible to maintain a reasonably constant virulence of the pathogen, large numbers of animals are required to achieve statistical significance. Therefore, the cost of utilising infectious models should be considered in addition to the previously mentioned ethical considerations. It would seem logical that quantitative correlations are obtained between resistance to experimental infections and functional immune changes so that the use of experimental infections can be limited as much as possible, if not totally.

- Biohazards related to the use of experimental infectious models are another critical issue. Microbial models should be selected taking into account the virulence of the pathogen for human beings, the possibility for adequate treatment, and the availability of vaccines to prevent the development of infectious diseases. Special confinement measures should also be endorsed to avoid contamination of animal facilities.

Experimental viral infections

Viral infections are particularly interesting to consider because they are major causes of morbidity and mortality in man. The outcome of the infection depends on the virus virulence, the susceptibility of the target organ and the host's immune competence.

A variety of models have been developed and proposed for non-clinical immunotoxicity evaluation (Kern, 1982). The most commonly used models will be briefly described.

Herpes virus infections

Infections due to the herpes virus family, including herpes simplex virus (HSV) and cytomegalovirus, are routinely used in non-clinical immunotoxicity evaluation. Most murine strains are not susceptible to HSV-1. The pathogenicity of HSV-2 depends on the animal's age and the route of inoculation. Cell-mediated immunity plays a pivotal role in the host's defence against HSV infection; macrophages also seem to be essential, which could explain the relative susceptibility of very young compared to adult animals. Natural killer cell activity was shown to play a major role in the defence againt cytomegalovirus (Selgrade *et al.*, 1992).

Influenza virus infections

Infections to influenza virus are also commonly used due to the availability of human isolates and the well-characterised mechanisms of pathogenesis (Lebrec and Burleson, 1994). Infections to influenza virus can be used in mice and rats. However, nasal administration induced mortality in mice, but not in rats. Quantification of infectious viral titre in the lungs of inoculated animals can also be used. Infections to influenza virus are helpful to assess the global consequences of depressed humoral immunity by drug treatment and chemical exposure.

Other viral infections

Infection due to encephalomyocarditis virus is caused by peritoneal or intranasal inoculation. Infection initially develops as viraemia followed by virus replication in the heart and the brain. Infected animals usually die within 6–8 days. Other viral infections include arboviruses, myxoviruses and retroviruses.

Bacterial infections

Many bacterial infectious models have also been developed.

Listeria monocytogenes *infections*

Infection due to *Listeria monocytogenes* is a prototypic infectious model (Bradley, 1995a). The main advantage of this model is the possible use of either rats or mice. Pathogens are cultured and a suspension of viable *Listeria* is quantified by turbidimetry and injected intravenously. The majority of injected *Listeria* are removed and destroyed by liver macrophages, whereas surviving *Listeria* multiply in the liver. In normal animals, the infection spontaneously stops within 5–6 days. This model is particularly interesting for investigation of cell-mediated immunity and macrophage functions. The final endpoint is either the death of infected animals, or the number of viable pathogens in the spleen and liver of inoculated animals, measured from the clearance of pathogens and/or organ bacterial colony counts.

Streptococcal infections

Streptococcal infections are another type of well-characterised and commonly used experimental infection models (Bradley, 1995b). *Streptococcus pneumoniae* and *Streptococcus zooepidemicus* are the two species widely utilised by immunotoxicologists. Complement and humoral immunity have been shown to play a major role in the defence against *Streptococcus pneumoniae*. Fresh inoculates of cultured bacteria are injected intravenously to mice. Mortality induced by inoculates of graded concentrations is noted shortly after injection. Humoral immunity is the major line of defence against *Streptococcus zooepidemicus* and the infection develops more slowly. Interestingly, rats are susceptible to *Streptococcus zooepidemicus* infections.

Other bacterial infections

Many bacterial infectious models have been proposed, mostly in mice. The most commonly used models include infections to *Escherichia coli*, *Salmonella typhimurium*, *Staphylococcus epidermidis*, and *Klebsiella pneumoniae*.

Parasitic infections

The main advantage of parasitic infections is the more consistent virulence of the microorganism compared to bacterial and viral pathogens, and the possibility of using other endpoints other than lethality.

Trichinella spiralis *infections*

Infection to *Trichinella spiralis* can be performed in both mice and rats (Dessein *et al.*, 1981; Van Loveren *et al.*, 1989, 1995). Cell-mediated immunity was shown to play a pivotal role in the response, whereas specific antibodies enhance the elimination of the parasite. The oral administration of larvae as a suspension induces an infection manifesting after one week with worms in the gut of 50 to 70 per cent of animals. Animals are killed after 14 days and adult worms are counted. Typically, immunosuppressed animals have a much higher number of encysted larvae than normal animals. Other endpoints can be measured, such as specific serum IgE levels.

Plasmodium *infections*

Infections to *Plasmodium berghei* and *Plasmodium yoelii* are very close to the human disease. Mice and certain strains of rats are given an intraperitoneal injection of the pathogens. Host resistance depends on specific antibodies. Evaluation criteria include mortality, erythrocyte count and haemoglobin level.

Toxoplasma gondii *infections*

Infection to *Toxoplasma gondii* is a murine model (Descotes *et al.*, 1991). The injection of a small (~100) number of parasites induces a progressively lethal disease (9–11 days). Because virulence is well preserved, results are reproducible and this is an adequate tool for investigation of host resistance following medicinal or chemical exposure.

Implanted tumours

Even though the concept of cancer immunosurveillance is no longer held as it was in the past, it is clear that the immune system plays a critical role in limiting the onset and development of neoplasias. Therefore, the exploration of antitumoural resistance has often been considered an important step in non-clinical immunotoxicity evaluation.

Due to their small incidence, spontaneous tumours cannot reasonably be used as cost-effective endpoints for evaluation of antitumoural resistance, and this is true even for spontaneous neoplasias with a greater incidence, such as leukaemias in Fischer 344 rats. Therefore, a number of implanted tumour models have been proposed (McKay, 1995). They include fibrosarcoma PYB6, sarcoma 1412, melanoma B16 or B16F10, tumoural ascitis EL-4, Lewis carcinoma in C57Bl/6 mice; tumoural ascitis L1210 or P388 in DBA/2 mice; MKSA (virus SV-40), TKL5, ascitis MOPC-104 in Balb/c mice; and C58NT or mammary adenoma MADB106 in rats.

General considerations

Every experimental tumour model comprises the same general steps: the injection of a suspension of tumour cells; a lag period of time to allow progression of the tumour; the onset of a palpable tumour; and finally the death of the animal.

Selection of the tumour model is largely dependent on the species and strain used, as the tumour must be syngeneic (to avoid specific rejection of the experimental tumour by the immune system). Tumour cells are prepared from solid tumours, cultured neoplastic cell lines, or ascites. Injection is performed either via the intradermal, subcutaneous, intratracheal or intramuscular route, or via the intraperitoneal route to induce ascites. Injection of a small volume, which induces tumours in 10 to 30 per cent of animals is preferred when an immunosuppressive chemical is under scrutiny. By contrast, a larger volume inducing tumours in 80 to 90 per cent of animals is recommended with immuno-stimulating chemicals.

Various evaluation criteria can be used: the incidence of tumours or the incidence of animals with one or more tumours, the period of time between the inoculation of tumour cells and the identification of palpable tumours, the time of tumour growth, the number of nodules, the death rate, or the delay of death. All these criteria have been more or less adequately standardised and validated.

Susceptibility assay towards tumour cells

This assay is performed as follows: 5×10^3 PYB6 or 10^6 4198T cells are injected sub-cutaneously two to three days after the last chemical administration to B6C3F1 mice. Ten to 20 per cent of control animals develop tumours. The animals are kept for 60 days and results are expressed as the number of animals with a palpable tumour, the delay between inoculation and tumour onset, and the mean volume of tumour nodules.

^{125}I-iododeoxyuridine (UdR) assay

The ^{125}I-iododeoxyuridine (UdR) assay is used to measure tumour mass in the lungs after administration of 1×10^5 B16F10 cells by intravenous injection or via inhalation. After 13–20 days, the animals are given an intravenous dose of 1 mg/kg fluorodeoxyuridine and one hour later 1 μCi of ^{125}I-UdR, which is readily incorporated into the DNA of cells with a short turnover. Mice are killed after 18 hours and the tumour mass is quantified from the radioactivity.

Other experimental diseases

Animal models of autoimmune diseases (Burkhardt and Kaldern, 1997; Theofilopoulos and Dixon, 1985) are commonly used by immunopharmacologists, but surprisingly draw limited attention for non-clinical immunotoxicity evaluation. This again presumably reflects the limited interest of immunotoxicologists on immunostimulation. Based on the clinical experience gained with therapeutic cytokines, animal models of autoimmune diseases could be proposed to fulfil the same role as experimental infections and implanted tumours for assessing the global consequences of immunostimulation associated with drug treatment or chemical exposure. Typically, these models include experimental allergic encephalomyelitis, uveoretinitis, and myasthenia. A number of models have been proposed in genetically defective mouse strains which spontaneously develop a disease with clinical and biological aspects similar to those seen in lupus erythematosus, such as NZB/NZW, BXB or MRL mice.

To test whether an immunotoxicant can interfere with *in vivo* allergic responses to an unrelated antigen, experimental models, such as passive cutaneous anaphylaxis in the rat or guinea-pig (Laschi-Loquerie *et al.*, 1984) are used.

References

BRADLEY, S.G. (1995a) *Listeria* host resistance model. In: *Methods in Immunotoxicology, vol. 2* (Burleson, G., Dean, J.H. and Munson, A.E., eds), pp. 169–179. New York: Wiley-Liss.

BRADLEY, S.G. (1995b) *Streptococcus* host resistance model. In: *Methods in Immunotoxicology, vol. 2* (Burleson, G., Dean, J.H. and Munson, A.E., eds), pp. 159–168. New York: Wiley-Liss.

BURKHARDT, H. and KALDERN, J.R. (1997) Animal models of autoimmune diseases. *Rheumatol. Int.*, **17**, 91–99.

DEAN, J.H., LUSTER, M.I., BOORMAN, G.A., LUEBKE, R.W. and LAUER, L.D. (1982) Application of tumor, bacterial, and parasite susceptibility assays to study immune alterations induced by environmental chemicals. *Environ. Health Perspect.*, **43**, 81–88.

DESCOTES, J., BROULAND, J.Ph., TEDONE, R. and VERDIER, F. (1991) Experimental toxoplasmosis in Swiss mice for assessing host resistance. *Toxicologist*, **11**, 207.

DESSEIN, A.J., PARKER, W.L., JAMES, S.L. and DAVID, J.R. (1981) IgE antibody and resistance to infection. I. Selective suppression of the IgE antibody response in rats diminishes the resistance and the eosinophil response to *Trichinella spiralis* infection. *J. Exp. Med.*, **153**, 423–436.

HERZYK, D.J., RUGGIERI, E.V., CUNNIGHAM, L., POLSKY, R., HEROLD, C., KLINKNER, *et al.* (1997) Single-organism model in host defense against infection: a novel immunotoxicologic approach to evaluate immunomodulatory drugs. *Toxicol. Pathol.*, **25**, 351–362.

KERN, E.R. (1982) Use of viral infections in animal models to assess changes in the immune system. *Environ. Health Perspect.*, **43**, 71–79.

LASCHI-LOQUERIE, A., DESCOTES, J., TACHON, P. and EVREUX, J.Cl. (1984) Influence of lead on hypersensitivity. Experimental study. *J. Immunopharmacol.*, **6**, 87–93.

LASCHI-LOQUERIE, A., EYRAUD, A., MORISSET, D., SANOU, A., TACHON, P., VEYSSEYRE, C. and DESCOTES, J. (1987) Influence of heavy metals on the resistance of mice toward infection. *Immunopharmacol. Immunotoxicol.*, **9**, 235–242.

LEBREC, H. and BURLESON, G.R. (1994) Influenza virus host resistance models in mice and rats: utilization for immune function assessment and immunotoxicology. *Toxicology*, **91**, 179–188.

LUSTER, M.I., PORTIER, C., PAIT, D.G., ROSENTHAL, G.J., GERMOLEC, D.R., CORSINI, E., *et al.*, (1994) Risk assessment in immunotoxicology. II. Relationships between immune and host resistance tests. *Fund. Appl. Toxicol.*, **21**, 71–82.

McKay, J.A. (1995) Syngeneic tumor cell models: B16F10 and PYB6. In: *Methods in Immunotoxicology, vol. 2* (Burleson, G., Dean, J.H. and Munson, A.E., eds), pp.143–157. New York: Wiley-Liss.

Selgrade, M.K., Daniels, M.J. and Dean, J.H. (1992) Correlation between chemical suppression of natural killer cell activity in mice and susceptibility to cytomegalovirus: rationale for applying murine cytomegalovirus as a host resistance model and for interpreting immunotoxicity testing in terms of risk of disease. *J. Toxicol. Environ. Health*, **37**, 123–137.

Theofilopoulos, A.N. and Dixon, F.J. (1985) Murine models of systemic lupus erythematosus. *Adv. Immunol.*, **37**, 269–390.

Thomas, P.T. and Sherwood, R.L. (1996) Host resistance models in immunotoxicology. In: *Experimental Immunotoxicology* (Smialowicz R.J. and Holsapple, M.P., eds), pp. 29–45. Boca Raton: CRC Press.

Van Loveren, H., Osterhaus, A.D.M.E., Nagel, J., Schuurman, H.J. and Vos, J.G. (1989) Quantification of IgA responses in the rat. Detection of IgA antibodies and enumeration of IgA antibody producing cells specific for ovalbumin and *Trichinella spiralis*. *Scand. J. Immunol.*, **28**, 377–381.

Van Loveren, H., Luebke, R.W. and Vos, J.G. (1995) Assessment of immunotoxicity with the parasitic infection model *Trichinella spiralis*. In: *Methods in Immunotoxicology, vol. 2* (Burleson, G., Dean, J.H. and Munson, A.E., eds), pp. 243–271. New York: Wiley-Liss.

14

Strategies for the Evaluation of Immunosuppression

Because immunosuppressive drugs were the first therapeutic agents with marked immuno-pharmacological activities to be introduced into the clinical setting, adverse health effects associated with immunosuppression have long been recognised, and hence unexpected immunosuppression associated with occupational and/or environmental chemical exposures has most often been the primary focus of immunotoxicologists. Even though hyper-sensitivity and autoimmunity are areas of increasing concern, immunosuppression remains a central aspect of non-clinical immunotoxicity evaluation, and indeed it is the only aspect of possible immunotoxic effects caused by medicinal products and chemicals, which can be adequately predicted today, at least to some extent.

The aim of this chapter is to provide an overview of possible, realistic and cost-effective strategies to be recommended for use in the prediction of unexpected immunosuppression, based on the author's own experience. Obviously, the final selection of strategies depends on the nature of the chemical to be tested as well as the expected modalities (magnitude, duration, etc.) of human exposure, but it should also take into account specific regulatory requirements, when available or applicable.

Tiered protocols

Following the pioneering work of American authors in the late 1970s, the concept of tiered (or multistep) protocols was introduced into the non-clinical evaluation of unex-pected immunosuppression associated with occupational and environmental chemicals (Dean *et al.*, 1979). This concept is now widely accepted and utilised, even though limited information is available regarding its predictive value and applicability to the non-clinical immunotoxicity evaluation of medicinal products, as few medicinal pro-ducts, such as morphine and methadone (De Waal *et al.*, 1998) have actually been evalu-ated using tiered protocols with the exception of several potent immunosuppressive agents, such as cyclosporin and azathioprine (De Waal *et al.*, 1995). Tiered protocols have even been suggested to be irrelevant for the immunotoxicity evaluation of novel medicinal products, in particular biotechnology-derived products (Cavagnaro *et al.*, 1995). Altern-atives, such as safety immunopharmacology studies (Choquet-Kastylevsky and Descotes,

1999), might prove useful as regards the evaluation of medicinal products, even though extensive efforts have to be paid to design, standardise and validate adequate animal models for use in safety immunopharmacology. Tiered protocols have nevertheless the key advantage of cost-effectiveness, as assays are performed step by step throughout the drug evaluation process, thus avoiding the possibility that poorly relevant assays are inappropriately performed.

Initially, tiered protocols included two tiers, but the use of three-tier protocols has more recently been advocated (Descotes *et al.*, 1993). Three-tier protocols seem more appropriate to take into account the current modalities of non-clinical toxicity testing and the present status of immunotoxicity evaluation in this context. Even though a few investigators proposed that several assays of immune function can be simultaneously performed in the same animal (Exon *et al.*, 1990), or that functional assays can be included in standard toxicity testing (Ladics *et al.*, 1995), the use of additional animals is generally recommended. This was a major limitation to the development of routine non-clinical immunotoxicity evaluation, as the use of additional animals necessarily results in extra cost. Non-functional tests can easily be included in conventional toxicity testing and one possibility is to include these tests in the proposed first tier, whereas functional and host resistance assays can be included in the second and third tiers, respectively. When a two-tier protocol is preferred, tier-1 and tier-2 of three-tier protocols are merged into the first tier.

Tier-1

Within the context of a three-tier protocol, the aim of the first tier is primarily the detection of changes suggestive or indicative of possible immunosuppressive effects. Non-functional assays are used and included in conventional repeated dose toxicity studies. In this context, this first tier is obviously, but also purposely limited to a few endpoints, which can be easily included in repeated dose toxicity studies. These endpoints, along with the routine clinical, haematological, biochemical and pathological endpoints of general toxicity studies, are expected to be helpful to trigger a warning signal that the tested xenobiotic might be immunotoxic. Based on the current scientific evidence, the understandable concern expressed by the public and the media, and finally the very limited extra cost, it is surprising that regulatory agencies throughout the world did not consider the systematic inclusion of such endpoints into conventional toxicity testing, whatever the chemical tested (in particular medicinal products), even though results cannot be expected to be more than gross warning signals requiring additional and careful assessment of potential immunotoxic effects.

At least three endpoints are proposed to be considered for inclusion into the first tier: the histopathological examination of major lymphoid organs, leukocyte (subset) analysis and the measurement of serum immunoglobulin levels.

Histopathological examination of major lymphoid organs

The histopathological examination of major lymphoid organs is generally considered of primary importance to detect unexpected immunosuppression associated with drug treatments or chemical exposures. As indicated in Chapter 9 of this volume, the few published immunotoxicity guidelines dealing with unexpected immunosuppression all include and highlight the histopathological examination of lymphoid organs. Lymphoid organs, which are recommended for study, include the thymus, the spleen, selected lymph nodes, the bone marrow and Peyer's patches. It should however be stressed that the significance of histological changes in Peyer's patches is ill-understood and it is unsure whether chemical substances consistently or reliably alter the architecture of Peyer's patches in relation to immunotoxicity despite theoretical considerations regarding the importance of mucosal immunity and Peyer's patches in the immune competence of mammals.

The determination of weight changes in lymphoid organs is commonly performed, but it is unknown whether, and in fact very unlikely that, mere increases or decreases in the weight of lymphoid organs are meaningful to the evaluation of immunotoxicity, unless weight change in a given lymphoid organ is associated with corresponding alterations of the histological architecture. Therefore, the determination of lymphoid organ weight is only to be considered as the initial step prior to histological examination.

The conventional histological examination of lymphoid organs is expected to provide clues to the detection of potential immunotoxicants. Histological examination of the lymphoid organs should be performed with care and skill, as laboratory artifacts and errors can easily occur. Atrophy or hyperplasia, either localised or generalised, are the major changes to be searched for. However, it is essential to keep in mind that histological changes provide at best limited information on the functional capacity of the immune system, so that histological changes can only be reliably interpreted in the light of immune function testing. Then, histological changes can be helpful and for instance, atrophy restricted to the T-independent, or conversely T-dependent, areas of lymph nodes and/or the spleen, can be useful to understand the mechanism of observed immunotoxic effects.

The value of conventional histological examination for predicting unexpected immunosuppression has been and is still debated. Although some investigators believe that conventional histology is sufficient to pick up unexpectedly immunosuppressive chemicals (Basketter *et al.*, 1994; Gopinath, 1996), others claim that conventional histology is not sensitive enough (De Waal *et al.*, 1996). An explanation for these conflicting views can probably be found in the lack of a clear and generally accepted definition of what immunosuppression actually is. When conventional histology is considered to be insufficient, two possibilities remain: either functional immune assays or improved histological techniques, often referred to as 'enhanced pathology'. As immunohistochemistry has developed so rapidly in recent years, new avenues of investigation have opened. However, even though it is now possible to identify and locate an enormous variety of tissue, cellular or subcellular components, and show their activation or resting status, it is still largely unproven whether and to what extent such sophisticated and expensive methods can indeed significantly increase the predictibility of conventional histopathological techniques. Efforts have certainly to be paid in this area to delineate the role of 'enhanced pathology' in the routine evaluation of immunotoxicity.

Leukocyte analysis

Haematology is a normal component of routine testing performed during every repeated administration toxicity study. Changes in white blood cell counts, such as lymphopenia, hyperleukocytosis, or hypereosinophilia, can be helpful to detect or suspect the immunotoxic effects of xenobiotics.

With the development of fluorescent activated cell sorter (FACS) and fluorescent cell counter techniques, it is now relatively easy to perform lymphocyte surface marker analysis in humans and most laboratory animal species. Quite often though, the analysis is restricted to B and T lymphocytes, as well as $CD4^+$ and $CD8^+$ T lymphocytes (Ladics *et al.*, 1994). It should be recalled that lymphocyte subset analysis is not a functional assay, and that the toxicological relevance of changes in lymphocyte subsets, e.g. a small decrease in the percentage of CD4[+] T lymphocytes, remains to be fully established.

However, a major advantage of lymphocyte subset analysis is that it can be performed in nearly every animal species commonly used in preclinical toxicity evaluation, so that comparison of animal and human data is readily possible. The predictive value of recently developed selective markers (Sopper *et al.*, 1997) remains largely unknown in the context of non-clinical immunotoxicity evaluation.

Serum immunoglobulin levels

The value of serum immunoglobulin levels is at best limited. However, because this is a very easy and inexpensive assay, this endpoint is widely used in the non-clinical immunotoxicological evaluation of xenobiotics, which obviously does not mean it is more likely to be reliable or even helpful. However, it is noteworthy that serum immunoglobulin levels proved to be one of the most sensitive endpoints to confirm the immunosuppressive properties of the therapeutic immunosuppressant azathioprine (Remandet *et al.*, 1992).

Tier-2

Funtional assays of the immune system are the core assays to be included in tier-2. When a two-tier protocol is preferred, functional assays are performed together with the non-functional assays described above. This tier is actually the first specific stage of non-clinical immunotoxicity evaluation. One major problem is that satellite animals are usually required because of sensitisation, which is suspected to induce misleading histological changes. From a theoretical viewpoint, four major aspects of the immune response, namely humoral immunity, cell-mediated immunity, phagocyte functions and NK cell activity, should be considered for study in this tier.

Humoral immunity

The humoral immune response is assessed in animals previously sensitised to a T-dependent antigen. The plaque-forming cell assay is, by far, the most widely used assay, and is the only assay which has been extensively validated in mice (Luster *et al.*, 1992) and also in rats (De Waal *et al.*, 1995; Richter-Reichhelm *et al.*, 1995; Van Loveren and Vos, 1989; White *et al.*, 1994).

Other techniques, such as ELISA and ELISPOT, are not considered to be reliable enough to be recommended today, a view however not held by all investigators (Kawabat, 1995; Temple *et al.*, 1993). In any case, comprehensive toxicological validation of these techniques remains to be conducted.

Cell-mediated immunity

The cell-mediated immune response has most commonly been explored using an *ex-vivo* lymphocyte proliferation assay: either the mitogen-induced proliferation assay or the mixed lymphocyte response assay (Smialowicz, 1995). Delayed-type hypersensitivity has long been considered to be less reliable, an assumption which was not confirmed by the NTP interlaboratory validation study (Luster *et al.*, 1992), so that either *ex-vivo* lymphocyte proliferation or delayed-type hypersensitivity can now be recommended with few, if any, objective data to select one assay instead of the other, with the exception of the investigator's experience and skill.

Phagocyte functions

A wide variety of assays can be used to assess phagocyte functions. In the context of non-clinical immunotoxicity evaluation, phagocytosis can be best assessed using a clearance assay, such as the *Listeria monocytogenes* clearance assay (Bradley, 1995), or a metabolic assay, such as the chemiluminescence assay (Verdier *et al.*, 1993), which can both be used in most laboratory animal species.

Due to the lack of adequate standardisation and toxicological validation, other test models to assess chemotaxis, adhesion or killing, are not recommended at this stage, and in fact, phagocyte function assays in general are inconsistently included in non-clinical immunotoxicity evaluation.

NK cell activity

The ^{51}Cr release assay is certainly the most widely used assay of NK cell activity (Djeu, 1995). Even though it has been widely used by immunologists, particularly in mice, and, to some extent, by toxicologists in rats, the repeatability and sensitivity are not optimal. Flow cytometry is expected to provide more reliable endpoints, but standardisation and validation studies have not yet been conducted.

Tier-3

This is by far the most critical and complex stage of non-clinical immunotoxicity evaluation, as the aim of this tier is both to confirm and reproduce changes observed in the course of previous tiers, and also more importantly, to assess the possible consequences of observed changes as well as to identify and understand the mechanisms involved for the purpose of immunotoxicity risk assessment.

Although no general rules are available, the conclusion as to whether a chemical is immunotoxic or not, largely depends on the results obtained with the selected assays. Close collaboration between immunologists, toxicologists and risk assessors is therefore essential to ensure that appropriate test models and relevant experimental conditions are selected, so that conclusions of the immunotoxicity risk assessment process are not flawed by excessive uncertainties.

Functional assays

Because functional assays are as cost-effectively limited as possible during the second tier, additional assays are often needed both to confirm and expand findings obtained in the course of the previous tiers. Alternative functional assays may be used, for instance ELISA, to evaluate humoral immunity when the plaque-forming cell assay has previously been used, or delayed-type hypersensitivity as an alternative to the lymphocyte proliferation assay.

The use of another antigen, such as keyhole limpet haemocyanin when sheep erythrocytes have been used in tier-2, also seems logical. The possibility that different immunological mechanisms may be involved in the response to different antigens should indeed be considered, as exemplified by delayed-type hypersensitivity responses (CD4[+] T lymphocytes are involved in the delayed-type hypersensivity response to sheep erythrocytes, whereas CD8[+] T lymphocytes are involved in the contact hypersensitivity response to chemical haptens). Similarly, it might prove useful to use an animal species in tier-3 different from that used in tier-2, for instance the mouse or a non-primate species, when the rat was used in tier-2.

Host resistance assays

The significance of observed functional immune changes cannot be better explored than by using host resistance assays, such as experimental infection models or, to a lesser extent, implanted tumours. It must be again emphasised that immune changes are not genuinely indicative of immunotoxicity, even though correlations have been demonstrated between some immune function changes and host resistance assays (Luster *et al.*, 1994; Selgrade *et al.*, 1992). Although experimental infection models may become increasingly difficult to perform in most Western countries, results obtained with these models are of critical relevance for identification of immunotoxicants. A careful selection of models based on the immune function changes previously shown to be induced by the tested chemical, and the known pathophysiology of the experimental infection, is absolutely essential to ensure that host resistance is adequately assessed.

Study of mechanism(s)

Emphasis should also be given to the study of mechanism(s) involved in the immunotoxic effects of xenobiotics. Mechanistic studies are essential for (immunotoxicity) risk assessment. As the variety of possible immunotoxic mechanisms is enormous, close collaboration between immunologists and toxicologists is again an absolute prerequisite to achieve the adequate selection of assays and the sound interpretation of results.

Implementation of tiered strategies

Even though there is wide consensus regarding the use of tiered protocols for non-clinical immunotoxicity evaluation, several possibilities exist, as already mentioned. One possibility is the use of two-tier protocols, in which non-functional assays are performed simultaneously with a limited battery of functional assays (tier-1), whereas extensive assessment of immune functions together with host resistance assays are including in the second tier. The possible design of three-tier protocols has been described above.

Interestingly, a third possibility was recently highlighted by the immunotoxicity testing guidelines for pesticides under the EPA's Toxic Substances Contact Act (*Federal Register*, 1997). The proposed protocol is in keeping with the concept of three-tier protocols, but restricts the battery of required immune function tests to a minimum. The conventional histological examination of lymphoid organs, which corresponds to tier-1, is performed during repeated administration toxicity studies. The plaque-forming cell assay is recommended as the pivotal immune function assay (tier-2), whereas additional assays, in particular NK cell activity and leukocyte subset analysis, are performed in a third tier, when the plaque-forming cell assay results in positive or doubtful findings.

Last but not least, a major issue regarding the implementation of non-clinical immunotoxicity evaluation strategies is the selection of chemicals which should be evaluated using such protocols. Importantly, the EPA immunotoxicity guidelines only considered pesticides, and the question arises as to whether the possible immunotoxicity of other xenobiotics, in particular medicinal products, should be similarly evaluated or whether additional assays should be considered.

Practical considerations

The practical conduct of the non-clinical immunotoxicity evaluation of a new chemical compound is not yet well codified, but a number of issues have to be addressed.

Selection of doses

As in any toxicity testing, the selection of doses is a critical step in non-clinical immuno-toxicity evaluation studies. As regards immunosuppression, at least one selected dose of the tested chemical should ideally induce significant functional changes in animals for extrapolation to man, but this is difficult for two main reasons: first, high doses may induce general toxic effects resulting in immune changes related to stress, decreased food intake or systemic toxicity, erroneously considered as directly immunotoxic; second, it is unclear what a toxicologically relevant immune change actually is.

Two different approaches can be proposed to address this issue. The first approach consists of the selection of several dose levels to show a typical dose-dependent response. The lower dose level is similar to the lower dose used in conventional short-term repeated administration toxicity studies. The next higher dose should not induce systemic toxicity and a decrease in body weight less than 10 per cent is usually considered adequate. A third still higher dose is sometimes recommended despite expected difficulties in the interpretation of results. A major issue related to this approach is the immunotoxicological relevance of induced histological and functional changes. In addition, no linear dose–response is often seen: one chemical can enhance the immune response at very low doses, exert no effects at higher dose levels, and finally be immunosuppressive at still higher dose levels, as exemplified with cadmium on the plaque-forming cell response of rats (Archimbaud *et al.*, 1992).

The second approach consists of the search for the highest 'immunologically safe' dose. Then, the high dose is the highest dose causing no functional changes. Although this approach has the advantage of avoiding indirect toxic effects on the immune system/response and the use of host resistance assays, a major limitation is the use of safety or uncertainty factors.

The use of either approach has consequences on the selection of assays. More robust assays are preferred with the first approach, whereas more sensitive assays seem more appropriate when selecting the second approach, because the demonstration of a lack of any functional change is critical in this context.

Duration of exposure

No generally accepted rules are available. The duration of exposure should be long enough to take into account the half-life of immunoglobulins, so that it is possible to show alterations in the synthesis of immunoglobulins. There is no evidence today that long-term studies are helpful to identify immunotoxic effects or changes which cannot be detected in short-term studies. However, the duration of exposure in most published studies was short-term, so it is premature to claim that long-term studies are unlikely to be of relevance.

The majority of immunotoxicity studies performed in rodents are 21–30-day studies, and only a few studies are 90-day studies. A shorter duration of exposure is not recommended, as suggested by some unexpected negative results of the NTP interlaboratory validation study in B6C3F1 mice with a 14-day duration of exposure (Luster *et al.*, 1992).

Route of exposure

The general rules of safety evaluation apply here, so that the route(s) of exposure should be the route(s) of expected exposure in humans. The oral route poses no particular difficulties as most immunotoxicity animal studies were performed following oral exposure. In contrast, the specificities of other routes, particularly inhalation, are not well established. Due to the lack of historical background data, the interpretation of results may be difficult.

Selection of species

The selection of species for non-clinical immunotoxicity evaluation has been a matter of much debate during the past two decades.

Mice versus rats

Initially, the mouse used to be the preferred species (Lebrec *et al.*, 1994; White *et al.*, 1985) because it was also the preferred species for immunological research, so that an enormous amount of information on the murine immune system and a nearly endless variety of immunological reagents were available. In contrast, limited information on the rat immune system and few immunological reagents were available. However, mice have limited applications during toxicity testing (essentially acute toxicity, selected genotoxicity assays, and carcinogenicity) so that the comparison of immunotoxicity changes in mice with general toxic effects is nearly, if not totally impossible. In contrast, rats are the primary species for toxicity studies so that the toxicity of nearly every compound (except therapeutic human proteins for instance) is investigated in rats. This proved to be the critical point, and rat studies have been increasingly performed in the 1980s, which expanded our knowledge on the rat immune system and facilitated improved availability of commercially available rat reagents (Van Loveren and Vos, 1989). Today, the rat is definitely the species of choice, which does not exclude the possibility of conducting immunotoxicity studies in other animal species.

Inbred versus outbred strains

Another matter of debate was whether an inbred or an outbred strain should be used for immunotoxicity evaluation. Initially, the B6C3F1 mouse was selected by the NTP, and by most American authors thereafter. An inbred or F1 strain was considered more appropriate to run immunological assays, as lesser interanimal variability was expected. When the rat began to be considered as a possible species for immunotoxicity evaluation, the Fischer F344 rat was proposed (White *et al.*, 1994). However, it is now largely recognised that outbred strains of rats, such as Sprague–Dawley or Wistar rats, can readily be used for immunotoxicity evaluation and that acceptable interanimal variability is noted (De Waal *et al.*, 1995; Richter-Reichhelm *et al.*, 1995).

Non-rodent species

Non-rodent species have uncommonly been used for non-clinical immunotoxicity evaluation, except non-human primates. Guinea-pigs, which have been a favoured species of immunologists for decades, have seldom been used (Burns *et al.*, 1996). Surprisingly, extremely few immunotoxicity dog studies have been published (Thiem *et al.*, 1988), although information on the dog immune system is available and the dog is the first non-rodent species for conventional toxicity testing. Non-human primates are more and more widely used, particularly for the non-clinical immunotoxicity evaluation of medicinal products.

Selection of assays

Several major issues can be identified when selecting assays to be included in a toxicity study, and this logically applies to immunotoxicity studies:

- *Standardization*. Efforts similar to those paid to the standardisation of clinical biochemistry tests have not been paid to immunology assays. Techniques are often markedly different between laboratories, even though they are located in the same city, the same university, or even the same building. The lack of standardised immunological reagents is a major difficulty. Reagents can be genuinely unstandardisable, such as foetal calf serum, but wide variability within the same batch of the same reagent, e.g. standard serum for immunoglobulin measurement, can unexpectedly be found.

- *Validation*. Immunotoxicologists devoted many efforts to the validation of assays and strategies, as repeatedly stressed in this volume. A fair number of interlaboratory validation studies have been performed: the US National Toxicology Program study in B6C3F1 mice (Luster *et al.*, 1992, 1994); the Fischer F344 cyclosporin A validation study which included nine laboratories from the USA and Europe (White *et al.*, 1994); the International Collaborative Immunotoxicity Study (ICIS) which comparatively investigated the immunotoxic effects of azathioprine and cyclosporin A in more than 20 laboratories worldwide under the auspices of the European Union and the International Programme for Chemical Safety (IPCS) (Dayan *et al.*, 1998); and the collaborative cyclosporin A study conducted by the German chemical and pharmaceutical industry (Richter-Reichhelm *et al.*, 1995) (a second model compound, hexachlorobenzene, is being investigated by an expanded panel of laboratories).

- *Significance of results*. The use of standardised and validated assays does not necessarily mean that biologically or clinically significant results will be obtained for the benefit of sound and reliable risk assessment. In this respect, the lack of human data is a major limitation to the interpretation of current immunotoxicity results.

Good Laboratory Practice

In the 1970s, the recognition of scientific frauds in data from non-clinical laboratory studies submitted to the US Food and Drug Administration resulted in the establishment of Good Laboratory Practice (GLP) which describes minimum requirements for testing facility management, data tracking, test article characterisation and handling, protocol development, data acquisition, and report preparation and keeping. GLP regulations have been issued throughout the world and although differences can be found between national and/or international GLP regulations, they all adhere to the same concepts.

The application of GLP rules to non-clinical immunotoxicity evaluation has seldom been addressed (Thomas *et al.*, 1995). Requirements, such as animal records, animal handling and housing, monitoring and maintenance of laboratory equipment, data acquisition and storage, or personnel training, have no specificities for immunotoxicity studies compared to other toxicity studies.

The standardisation of reagents and assays is a critical issue. Most immunological reagents have not been as extensively standardised as other types of reagents used in non-clinical toxicity studies. It is unsure whether the careful labelling of reagents, identification of commercial sources, conditions of storage and handling, are sufficient to ensure reproducible quality. Standard operating procedures can be established and strictly adhered to by authorised personnel, but the lack of inter-laboratory standardisation can result in poorly reproducible results. Although the lack of adequate standardisation is not intrinsically an issue related to GLP regulations, this should be kept in mind and efforts should be paid by laboratory immunologists, reagent manufacturers and immunotoxicologists to ensure a long-awaited better quality of procedures.

References

ARCHIMBAUD, Y., VERDIER, F., HENGE-NAPOLI, M.H. and DESCOTES, J. (1992) Changes in plaque forming cells and phagocytosis of rats subchronically exposed to low doses of cadmium. *J. Toxicol. Occup. Environ. Health*, **1**, 24–28.

BASKETTER, D.A., BREMMER, J.N., KAMMÜLLER, M.E., KAWABATA, T., KIMBER, I., LOVELESS, S.E., *et al.* (1994) The identification of chemicals with sensitising or immunosuppressive properties in routine toxicology. *Food Chem. Toxicol.*, **26**, 527–539.

BRADLEY, S.G. (1995) *Listeria* host resistance model. In: *Methods in Immunotoxicology, vol. 2* (Burleson, G., Dean, J.H. and Munson, A.E., eds), pp. 169–179. New York: Wiley-Liss.

BURNS, L.A., DUWE, R.L., JOVANOVIC, M.L., SEATON, T.D., JEAN, P.A., GALLAVAN, R.H., *et al.* (1996) Development and validation of the antibody-forming cell response as an immunotoxicological endpoint in the guinea-pig. *Toxicol. Meth.*, **6**, 193–212.

CAVAGNARO, J.A., MIELACH, F.A. and MYERS, M.J. (1995) Perspectives on the immunotoxicological evaluations of therapeutic products: assessement of safety. In: *Methods in Immunotoxicology, vol. 1* (Burleson, G., Dean, J.H. and Munson, A.E., eds), pp. 37–50. New York: Wiley-Liss.

CHOQUET-KASTYLEVSKY, G. and DESCOTES, J. (1999) Safety immunopharmacology. In: *The Handbook of Safety Pharmacology* (Descotes, J., ed.) in press. London: Taylor & Francis.

DAYAN, A.D., KUPER, F., MADSEN, C., SMIALOWICZ, R.J., SMITH, E., VAN LOVEREN, H., *et al.* (1998) Report of interlaboratory validation study of assessment of direct immunotoxicity in the rat. *Toxicology*, **125**, 183–201.

DEAN, J.H., PADARASINGH, N.L. and JERRELLS, T.R. (1979) Assessment of immunobiological effects induced by chemicals, drugs or food additives. I. Tier testing and screening approach. *Drug Chem. Toxicol.*, **2**, 5–17.

DESCOTES, J., VIAL, T. and VERDIER, F. (1993) The how, why and when of immunological testing. *Comp. Haematol. Intern.*, **3**, 63–66.

DE WAAL, E.J., TIMMERMAN, H.H., DORTANT, P.M., KRAJNC, M.A.M. and VAN LOVEREN, H. (1995) Investigation of a screening battery for immunotoxicity of pharmaceuticals within a 28-day oral toxicity study using azathioprine and cyclosporin A as model compounds. *Regul. Toxicol. Pharmacol.*, **21**, 327–338.

DE WAAL, E.J., VAN DER LAAN, J.W. and VAN LOVEREN, H. (1996) Immunotoxicity of pharmaceuticals: a regulatory perspective. *Toxicol. Ecotoxicol. News*, **3**, 165–172.

DE WAAL, E.J., VAN DER LAAN, J.W. and VAN LOVEREN, H. (1998) Effects of prolonged exposure to morphine and methadone on in vivo parameters of immune function in rats. *Toxicology*, in press.

DJEU, J.Y. (1995) Natural killer activity. In: *Methods in Immunotoxicology, vol. 1* (Burleson, G., Dean, J.H. and Munson, A.E., eds), pp. 437–449. New York: Wiley-Liss.

EXON, J.H., BUSSIERE, J.L. and MATHER, G.G. (1990) Immunotoxicity testing in the rat: an improved multiple assay model. *Int. J. Immunopharmac.*, **12**, 699–701.

Federal Register (1997) Toxic Substances Control Act Test Guidelines. Final Rule. **62**, 43819–43864.

GOPINATH, C. (1996) Pathology of toxic effects on the immune system. *Inflamm. Res.*, **2**, S74–S78.

KAWABAT, T.T. (1995) Enumeration of antigen-specific antibody-forming cells by the enzyme-linked immunospot (ELISPOT) assay. In: *Methods in Immunotoxicology, vol. 1* (Burleson, G., Dean, J.H. and Munson, A.E., eds), pp. 125–135. New York: Wiley-Liss.

LADICS, G.S. and LOVELESS, S.E. (1994) Cell surface marker analysis of splenic lymphocyte populations of the CD rat for use in immunotoxicological studies. *Toxicol. Meth.*, **4**, 77–91.

LADICS, G.S., SMITH, C., HEAPS, K., ELLIOTT, G.S., SLONE, T.W. and LOVELESS, S.E. (1995) Possible incorporation of an immunotoxicological functional assay for assessing humoral immunity for hazard identification purposes in rats on standard toxicology study. *Toxicology*, **96**, 225–238.

LEBREC, H., BLOT, C., PEQUET, S., ROGER, R., BOHUON, C. and PALLARDY, M. (1994) Immunotoxicological investigation using pharmaceutical drugs: in vivo evaluation of immune effects. *Fund. Appl. Toxicol.*, **23**, 159–168.

LUSTER, M.I., PORTIER, C., PAIT, D.G., WHITE, K.L., GENNINGS, C., MUNSON, A.E. and ROSENTHAL, G.J. (1992) Risk assessment in immunotoxicology. I. Sensitivity and predictability of immune tests. *Fund. Appl. Toxicol.*, **18**, 200–210.

LUSTER, M.I., PORTIER, C., PAIT, D.G., ROSENTHAL, G.J., GERMOLEC, D.R., CORSINI, E., et al. (1994) Risk assessment in immunotoxicology. II. Relationships between immune and host resistance tests. *Fund. Appl. Toxicol.*, **21**, 71–82.

REMANDET, B., BALL, D., FOURCINE, N., PFERSDORFF, C., GOUY, D. and DESCOTES, J. (1992) Analysis of total and specific IgG antibody responses: value in immunotoxicological assessment. *Toxicol. Letters*, **suppl.1**, 146 (abstract).

RICHTER-REICHHELM, H.-B., DASENBROCK, C.A., DESCOTES, G., EMMENDÖRFER, H.U.E., HARLEMANN, J.H., HILDEBRAND, B., et al. (1995) Validation of a modified 28-day rat study to evidence effects of test compounds on the immune system. *Regul. Toxicol. Pharmacol.*, **22**, 54–56.

SELGRADE, M.K., DANIELS, M.J. and DEAN, J.H. (1992) Correlation between chemical suppression of natural killer cell activity in mice and susceptibility to cytomegalovirus: rationale for applying murine cytomegalovirus as a host resistance model and for interpreting immunotoxicity testing in terms of risk of disease. *J. Toxicol. Environ. Health*, **37**, 123–137.

SMIALOWICZ, R.J. (1995) *In vitro* lymphocyte proliferation assays: the mitogen-stimulated response and the mixed-lymphocyte reaction in immunotoxicity testing. In: *Methods in Immunotoxicology, vol. 1* (Burleson, G., Dean, J.H. and Munson, A.E., eds), pp. 197–210. New York: Wiley-Liss.

SOPPER, S., STAHL-HENNIG, C., DEMUTH, M., JOHNSTON, I.C., DORRIES, R. and TER-MEULEN, V. (1997) Lymphocyte subsets and expression of differentiation markers in blood and lymphoid organs of rhesus monkeys. *Cytometry*, **29**, 351–352.

TEMPLE, L., KAWABATA, T.T., MUNSON, A.E. and WHITE, K.L. (1993) Comparison of ELISA and plaque-forming cell assays for measuring the humoral response to SRBC in rats and mice treated with benzo[a]pyrene or cyclophosphamide. *Fund. Appl. Toxicol.*, **21**, 412–419.

THIEM, P.A., HALPER, L.K. and BLOOM, J.C. (1988) Techniques for assessing canine mononuclear phagocyte function as part of an immunotoxicologic evaluation. *Int. J. Immunopharmac.*, **10**, 765–771.

THOMAS, P.T., BOYNE, R.A. and SHERWOOD, R.L. (1995) Good Laboratory Practice considerations: nonclinical immunotoxicology studies. In: *Methods in Immunotoxicology, vol. 1* (Burleson, G.R., Dean, J.H. and Munson, A.E., eds), pp. 25–36. New York: Wiley-Liss.

VAN LOVEREN, H. and VOS, J.G. (1989) Immunotoxicological considerations: a practical approach to immunotoxicity testing in the rat. In: *Advances in Applied Toxicology* (Dayan, A.D. and Payne, A.J., eds), pp. 143–165. London: Taylor & Francis.

VERDIER, F., CONDEVAUX, F., TEDONE, R., VIRAT, M. and DESCOTES, J. (1993) In vitro assessment of phagocytosis. Interspecies comparison of chemiluminescence response. *Toxicol. in vitro*, **7**, 317–320.

WHITE, K.L., SANDERS, V.M., BARNES, D.W., SHOPP, G.M. and MUNSON, A.E. (1985) Immunotoxicological investigations in the mouse: general approach and methods. *Drug Chem. Toxicol.*, **8**, 299–331.

WHITE, K.L., GRENNINGS, C., MURRAY, M.J. and DEAN, J.H. (1994) Summary of an international methods validation study, carried out in nine laboratories, on the immunological assessment of cyclosporin A in the Fischer 344 rat. *Toxicol. in vitro*, **8**, 957–961.

15

Identification of Chemical Sensitisers

Hypersensitivity (allergic) reactions are among the most frequent adverse effects of medicinal products in human beings. Because of the occurrence of excessively frequent and/or severe allergic reactions in relation to the expected therapeutic benefit, a number of drugs have been withdrawn from the market in the past two decades and this has had obvious negative consequences from a medical as well as industrial perspective. Similarly, hypersensitivity reactions due to industrial chemicals, pesticides, food additives, or cosmetics are relatively common and a matter of growing concern.

In spite of this, it is extremely difficult today to predict whether a new chemical entity, and in particular a novel medicinal product, is likely to be a sensitising agent in man (Choquet-Kastylevsky and Descotes, 1998). Explanations can be offered to account for this situation:

- Until recently, immune allergic reactions induced by medicinal products and other xenobiotics were considered unpredictable adverse effects. Based on this assumption, very limited efforts were being paid to design useful predictive animal or *in vitro* models and also, at least to some extent, to expand our limited understanding of fundamental causative mechanisms.

- The vast majority of medicinal products are small-molecular-weight substances with no or limited chemical reactivity, so that they cannot easily play the role of haptens in contrast to chemically reactive compounds, or be directly immunogenic, as are large macromolecules, such as microbial extracts and proteins. No models are available which can reproduce the formation of haptens from the administered medicinal product, either *in vivo* or *in vitro*, in realistic conditions.

- Because of these limitations, most research works focused on specific situations and chemicals so that few data enable extrapolation of results to other situations or chemicals.

Anaphylaxis models

Anaphylactic reactions involve reaginic antibodies, which are IgE in man, IgE and IgG_1 in the guinea-pig, and IgE and IgG_{2a} in the rat or the mouse. These antibodies bind to high-affinity receptors on target cells, i.e. mast cells and basophils. The resulting specific antigen–antibody reaction triggers the release of preformed mediators, such as histamine, and the subsequent synthesis of phospholipid-derived mediators, such as the prostaglandins and leukotrienes. Anaphylaxis models which attempt to reproduce this phenomenon experimentally, include systemic and local anaphylaxis models.

Systemic anaphylaxis models

Systemic anaphylaxis models are usually performed in the guinea-pig (Verdier *et al.*, 1994), and rarely in the mouse. Rat models, sometimes considered to be more appropriate models of human systemic anaphylaxis, are more difficult to perform and therefore their use is mostly restricted to pharmacological investigations (Ufkes *et al.*, 1983). A major and often debated issue with current models is the different susceptibility of animal species, in particular guinea-pigs, towards anaphylaxis.

Whatever the model, the experimental design consistently includes three phases, namely the sensitising phase, the rest phase and the eliciting phase. Sensitisation is tentatively induced by one, or more commonly several injections of the antigen via the subcutaneous, intradermal or intramuscular route, with a time interval of one to several days between each injection. The first sensitising injection may be combined with the injection of aluminium hydroxide as an adjuvant, either as a mixture or at a different site, to increase the production of IgE (Levine and Vaz, 1970). The use of complete Freund's adjuvant is not recommended as IgE production is not increased or even decreased after injection of complete Freund's adjuvant (Levine *et al.*, 1971). The rest period between the termination of the sensitising phase and the eliciting injection is generally 14–21 days. The eliciting injection is usually given intravenously. When antigen-specific reaginic antibodies are present, this results in an antigen–antibody reaction causing the death of the animal, or alternatively in clinical symptoms of gradable severity, such as nose licking or rubbing, weakened muscle tone, prostration, cyanosis, respiratory disorders and convulsions.

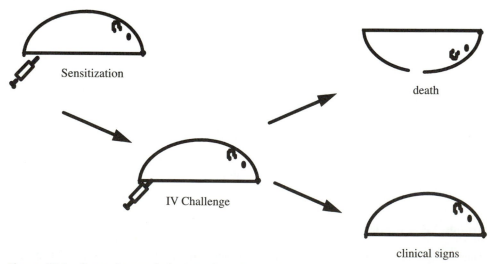

Figure 15.1   Systemic anaphylaxis in the guinea-pig

Despite the frequent utilisation of systemic anaphylaxis models in the guinea-pig for the past 30 years, in particular in the pharmaceutical industry, limited efforts have actually been paid to the standardisation and validation of these models. The selection of experimental conditions, such as the sensitising and eliciting doses, the number of sensitising injections, the route of optimal relevance for sensitisation and elicitation, or the duration of the rest period, is therefore generally based on the investigator's experience instead of objective data. Reproducible results have however been obtained: a panel of commonly used human vaccines, several non-human proteins and microbial extracts induced systemic anaphylactic responses in guinea-pigs, a finding in agreement with the clinical experience available with these compounds. False positive responses were nevertheless seen in guinea-pigs injected with human or humanised proteins, such as human immunoglobulins, recombinant IL-2 and human albumin, emphasising the total lack of relevance of guinea-pig systemic anaphylaxis models for the evaluation of humanised or human compounds (Brutzkus *et al.*, 1997).

In fact, the major current limitation to the use of guinea-pig systemic anaphylaxis models is the failure to induce anaphylactic response in animals injected with small-molecular-weight medicinal products with limited or no chemical reactivity, such as hydroxocobalamin and procaine (Chazal *et al.*, 1994), which indicates that such models are unable to reproduce experimentally the formation of reactive intermediates from parent molecules to play the role of haptens. However, when small-molecular-weight products of sufficient chemical reactivity, such as benzylpenicillin (Muranake *et al.*, 1978), are tested, a positive anaphylactic response may be seen.

Guinea-pig models are also commonly used for the detection of respiratory allergens (Botham *et al.*, 1988, 1989; Kimber and Dearman, 1997; Pauluhn and Eben, 1991; Sarlo and Clark, 1992; Sarlo and Karol, 1994). Respiratory sensitisation resulting in immediate reactions can be investigated following animal exposure to atmospheres of free and protein-bound allergens and subsequent inhalation challenge with the same form of the allergen, or following either topical, subcutaneous, intradermal, or intratracheal exposure to the free chemical, and inhalation challenge with the free or protein-bound chemical. Guinea-pig models proved helpful to predict which proteins and highly reactive low-molecular-weight chemicals are likely to be respiratory allergens, but the same limitations exist as regards their predictibility for non- or poorly reactive small molecules. Models for the assessment of respiratory sensitisers in other animal species, such as the mouse and the rat, have also been proposed. They grossly rely on the same experimental conditions and have as yet no proven superiority over guinea-pig models.

Local anaphylaxis models

Passive cutaneous anaphylaxis was designed in the early 1950s (Ovary, 1951a, 1951b) and since then has been the most widely used anaphylaxis model in the guinea-pig, the rat and the mouse, although other animal species, such as the monkey and the dog can also be utilised (Watanabe and Ovary, 1977).

Briefly, mice or guinea-pigs receive several intradermal or subcutaneous injections of the tested chemical (with or without the addition of an adjuvant, such as aluminium hydroxide or *Bordetella pertussis*) at intervals of one to several days in an attempt to induce sensitisation. The serum from previously (and supposedly) sensitised animals is injected into the dermis of naive animals, e.g. guinea-pigs, rats or mice. The tested chemical mixed to a dye, such as Evans blue, is injected intravenously. Because of the increased capillary permeability due to the local antigen–antibody reaction, extravasation of the dye is seen with the subsequent formation of a blue spot, the diameter of which can be measured. This is the so-called direct passive cutaneous anaphylaxis. In reverse passive cutaneous anaphylaxis, the tested chemical is injected locally and the serum from supposedly sensitised animals is injected intravenously.

Despite the extremely wide use of passive cutaneous anaphylaxis models, no validation has ever been conducted. In a comparison of systemic anaphylaxis with passive cutaneous anaphylaxis using a limited panel of reference compounds, passive cutaneous anaphylaxis in the guinea-pig was not found to be more sensitive than systemic anaphylaxis (Verdier *et al.*, 1994).

Contact sensitisation assays

Although contact sensitisation is definitely an immunotoxic effect of medicinal products and other xenobiotics, contact sensitisation assays have rarely been considered as part of the non-clinical immunotoxicity evaluation until recently.

Following the pioneering work of Draize (Draize *et al.*, 1944), the guinea-pig became the species of choice for contact sensitisation assays, particularly after refined technical procedures were proposed, first by Buehler (1965), then Magnusson and Kligman (1969). Despite the enormous number of published results, contact sensitisation assays in the guinea-pig present some limitations and efforts have been paid to the development of mouse assays during the past decade.

Guinea-pig contact sensitisation assays

A number of guinea-pig assays and modifications have been proposed, but none conclusively proved the most useful (Guillot *et al.*, 1983). However, guinea-pig test methods using complete Freund's adjuvant have been shown (Marzulli and Maguire, 1982) and are generally considered, although not undebatingly, to be superior in determining skin sensitisers. In fact, the technical skill and experience of the investigator are critical for the quality and the reliability of results. As the Magnusson and Kligman test and the Buehler test have been more widely used than any other guinea-pig assay, a short description of these tests is provided. Detailed information on other guinea-pig assays has been reviewed (Andersen and Maibach, 1985; Kimber and Maurer, 1996; Klecak, 1996).

Guinea-pig maximisation test

This test was initially described by Magnusson and Kligman (1969). The tested chemical is generally studied in 20 guinea-pigs, and 10 or 20 additional animals are used as controls. The induction phase includes three paired intradermal injections (0.1 ml each) of complete Freund's adjuvant, the tested chemical diluted in an adequate vehicle, such as water, paraffin oil or propylene glycol, and a mixture of the tested chemical in complete Freund's adjuvant (1:1). At day +7, the tested chemical is applied through an occlusive patch on the shaved scapular area for 48 hours. The elicitation is performed at day +21 on the shaved flank by the application of an occlusive patch for 24 hours. The skin reaction is read after 24 and 48 hours. Control animals receive the vehicle during the sensitisation phase and the tested chemical during the eliciting phase. Results are evaluated using a semi-quantitative rating scale: 0 = no reaction; 1 = scattered mild

Table 15.1 Contact allergenicity ranking according to
Magnusson and Kligman (1969)

Sensitisation rate (%)	Grade	Classification
0–8	I	Weak
9–28	II	Mild
29–64	III	Moderate
65–80	IV	Strong
81–100	V	Extreme

redness; 2 = moderate and diffuse redness; 3 = intense redness and swelling. Modifications of the rating scale have been proposed and recommended by several authors and regulatory agencies (Schlede and Eppler, 1995). Finally, the number of sensitised animals in the test group is an indication of the potency of the contact allergen (see Table 15.1).

The Magnusson and Kligman test is considered to be a very sensitive test, but with a tendency to exaggerate the response to weak human sensitisers. It is generally the reference method for the assessment of chemical contact sensitisers, particularly in Europe.

Buehler test

In the Buehler test (Buehler 1965, 1994), the induction phase consists of one to three epicutaneous applications under occlusion of a slighty irritating concentration of the tested chemical. The skin area in the left scapular area, previously shaved, is approximately 8 cm². Occlusive patches are maintained for six hours. After a rest period of 14 days, the elicitation is performed by topical application of 0.2–0.5 ml of a non-irritating concentration of the tested chemical, under occlusion maintained for six hours. The skin reaction is read after 24 and 48 hours.

This test is generally considered less sensitive and is also more time-consuming. It is often the preferred method in the USA, in particular for testing of commercial products using the concentration of the final formulation.

Murine assays

Although guinea-pig models are well standardised and have been extensively validated, a number of limitations can be found regarding the housing of animals, the susceptibility of guinea-pigs to infectious diseases, the amount of tested chemical required to perform guinea-pig tests, and the relative inability of these models to detect weak sensitisers in humans.

Surprisingly, the mouse has long been considered unable to mount a delayed-type hypersensitivity response (Crowle and Crowle, 1960). A critical advance was the use of ear swelling as a component of the murine delayed-type hypersensitivity response (Asherson and Ptak, 1968). It is actually very easy to induce ear swelling, evidenced by increased ear thickness following application of the contact sensitiser on the ear after prior sensitisation by several topical applications either on the shaved abdomen or on the opposite ear.

Mouse ear swelling test (MEST)

The MEST was initially proposed in 1986 (Gad *et al.*, 1986) and the technique later refined (Gad, 1994). CF-1 or Balb/c female mice are given 250 IU/g body weight of vitamin A. At day +1, mice receive two injections of 20 µl complete Freund's adjuvant, and 100 µl of the tested chemical at a slightly irritating concentration is applied on the shaved abdomen at days +1, +2, +4, and +6. At day +11, 20 µl of a non-irritating concentration of the tested chemical is applied on the ear and ear thickness is measured using a dial caliper, immediately before ear application, and after 24 and 48 hours.

This test which was initially considered very promising, failed to prove superior to guinea-pig assays. It is actually rather less sensitive, and as time-consuming and expensive as most guinea-pig assays. One may even wonder whether the MEST is likely to be recommended in the near future.

Other murine tests based on the same general principles have been proposed, namely the vitamin-A enriched ear swelling test (Maisey and Miller, 1984) and the mouse ear sensitisation assay (Descotes, 1988). They did not generate more valid results than the MEST.

Local lymph node assay (LLNA)

The local lymph node assay (LLNA), formerly called murine or auricular local lymph node assay is an original and attractive method (Kimber and Weisenberger, 1989; Kimber *et al.*, 1994). CBA/Ca mice receive topical applications of the tested chemical on the dorsal side of one ear for three consecutive days. After a rest period of five days, mice are injected intravenously with 20 µCi of tritiated thymidine in 250 µl of PBS. Five hours later, mice are killed, their auricular lymph nodes are removed and the lymph node lymphocytes cultured. The contact sensitising potential is assessed from the thymidine incorporation index. The positive threshhold was initially set to be 3, and subsequent validation studies largely confirmed that this threshhold was appropriate.

Interlaboratory validation studies showed that the LLNA can generate reproducible and consistent results (Kimber *et al.*, 1995). In addition, comparative studies showed similar results with the LLNA in mice and the maximisation test in guinea-pigs (Basketter and Scholes, 1992; Basketter *et al.*, 1993).

Evaluation strategies

The limited number of assays which have been properly standardised and validated is a major problem when establishing a strategy for the preclinical evaluation of the sensitising potential of xenobiotics.

Systemic and local anaphylaxis models are felt appropriate for the study of large-molecular-weight molecules, such as proteins, peptides, vaccines, and microbial extracts, provided they are not humanised or of human origin. In contrast, these models are not applicable to low-molecular-weight molecules which play the role of haptens. Coupling a small molecule to a macromolecule carrier has no predictive value since antibodies are always raised against the coupled chemical. This approach may however be helpful when the risk for cross-reactivity between two chemicals needs to be investigated.

Contact sensitisation models can also be used to predict the overall sensitising potential of xenobiotics: a parallelism was noted between the contact sensitising potential of 59/70 medicinal products and other xenobiotics in the guinea-pig, and the rate of allergic reactions reported in humans exposed to these compounds either by the oral route or by inhalation (Vial and Descotes, 1994). Combining contact sensitisation *in vivo* and an *in vitro* assay, such as the macrophage migration inhibition test, was shown to result in enhanced predictivity of the risk for systemic allergic reactions (Laschi-Loquerie *et al.*,

1987). Although these findings are empirical, they probably reflect the similarities of antigen presentation and immunological reactivity following chemical exposure whatever the route.

The prediction of allergic reactions induced by medicinal products and other xenobiotics is therefore extremely difficult and no general approach can be recommended. This is particularly true for humanised biotechnology-derived products and food allergens.

References

ANDERSEN, K.E. and MAIBACH, H.I. (1985) Guinea pig allergy tests: an overview. *Toxicol. Indust. Health*, **1**, 43–66.

ASHERSON, G.L. and PTAK, W. (1968) Contact and delayed hypersensitivity in the mouse. I. Active sensitization and passive transfer. *Immunology*, **15**, 405–416.

BASKETTER, D.A. and SCHOLES, E.W. (1992) A comparison of the local lymph node assay with the guinea pig maximisation test for the detection of a range of contact allergens. *Fd Chem. Toxicol.*, **30**, 63–67.

BASKETTER, D.A., SELBIE, E., SCHOLES, E.W., LEES, D., KIMBER, I. and BOTHAM, P.A. (1993) Results with OECD recommended positive control sensitizers in the maximisation, Buehler and local lymph node assays. *Fd Chem. Toxicol.*, **31**, 63–67.

BOTHAM, P.A., HEXT, P.M., RATTRAY, N.J., WALSH, S.T. and WOODCOCK, D.R. (1988) Sensitization of guinea pigs by inhalation exposure to low molecular weight chemicals. *Toxicol. Letters*, **41**, 159–173.

BOTHAM, P.A., RATTRAY, N.J., WOODCOCK, D.R., WALSH, S.T. and HEXT, P.M. (1989) The induction of respiratory allergy in guinea pigs following intradermal injection of TMA: a comparison with the response to DNCB. *Toxicol. Letters*, **47**, 25–39.

BRUTZKUS, B., COQUET, B., DANVE, B. and DESCOTES, J. (1997) Systemic anaphylaxis in guinea-pigs: intra-laboratory validation study. *Fund. Appl. Toxicol.*, **36**, suppl., 192.

BUEHLER, E.V. (1965) Delayed contact hypersensitivity in the guinea pig. *Arch. Dermatol.*, **91**, 171–177.

BUEHLER, E.V. (1994) Occlusive patch method for skin sensitization in guinea pigs: the Buehler method. *Fd Chem. Toxicol.*, **32**, 97–101.

CHAZAL, I., VERDIER, F., VIRAT, M. and DESCOTES, J. (1994) Prediction of drug-induced immediate hypersensitivity in guinea-pigs. *Toxicol. in vitro*, **8**, 1045–1049.

CHOQUET-KASTYLEVSKY, G. and DESCOTES, J. (1998) Predictive models of allergy. *Toxicology*, **129**, 27–35.

CROWLE, A.J. and CROWLE, C.M. (1960) Contact sensitivity in mice. *J. Allergy*, **32**, 302–320.

DESCOTES, J. (1988) Identification of contact allergens: the mouse ear sensitization assay. *J. Toxicol. Cutan. Ocular. Toxicol.*, **74**, 263–272.

DRAIZE, J.H., WOODWARD, G. and CALVERY, H.O. (1944) Methods for the study of irritation and toxicity of substances applied topically to the skin and mucous membranes. *J. Pharmacol. Exp. Ther.*, **82**, 377–390.

GAD, S.C. (1994) The mouse ear swelling test (MEST) in the 1990s. *Toxicology*, **93**, 33–46.

GAD, S.C., DUNN, B.J., DOBBS, D.W., REILLY, C. and WALSH, R.D. (1986) Development and validation of an alternative dermal sensitization test: the mouse ear swelling test (MEST). *Toxicol. Appl. Pharmacol.*, **84**, 361–368.

GUILLOT, J.P., GONNET, J.F., CLEMENT, C. and FACCINI, J.M. (1983) Comparison study of methods chosen by the Association Française de Normalisation (AFNOR) for evaluating sensitizing potential in the albino guinea-pig. *Fd Chem. Toxicol.*, **21**, 795–805.

KIMBER, I. and DEARMAN, R.J. (1997) *Toxicology of Chemical Respiratory Hypersensitivity*. London: Taylor & Francis.

KIMBER, I. and MAURER, T. (1996) *Toxicology of Contact Hypersensitivity*. London: Taylor & Francis.

KIMBER, I. and WEISENBERGER, C. (1989) A murine local lymph node assay for the identification of contact allergens. *Arch. Toxicol.*, **63**, 274–282.

KIMBER, I., DEARMAN, R.J., SCHOLES, E.W. and BASKETTER, D.A. (1994) The local lymph node assay: developments and applications. *Toxicology*, **93**, 13–31.

KIMBER, I., HILTON, J., DEARMAN, R.J., GERBERICK, G.F., RYAN, C.A., BASKETTER, D.A., *et al.* (1995) An international evaluation of the local lymph node assay and comparison of modified procedures. *Toxicology*, **103**, 63–73.

KLECAK, G. (1996) Test methods for allergic contact dermatitis in animals. In: *Dermatotoxicology*, 5th edition (Marzulli, F.N. and Maibach, H.I., eds), pp. 437–460. London: Taylor & Francis.

LASCHI-LOQUERIE, A., TACHON, P., VEYSSEYRE, C. and DESCOTES, J. (1987) Macrophage migration inhibition test to evaluate the sensitizing potential of drugs in the guinea pig. *Arch. Toxicol.*, **suppl. 11**, 325–328.

LEVINE, B.B. and VAZ, N.M. (1970) Effect of combinations of inbred strain, antigen and antigen dose on immune responsiveness and reagin production in the mouse. *Int. Archs. Allergy*, **39**, 156–171.

LEVINE, B.B., CHANG, H. and VAZ, N.M. (1971) The production of hapten-specific reaginic antibodies in the guinea pig. *J. Immunol.*, **104**, 29–33.

MAGNUSSON, B. and KLIGMAN, A.M. (1969) The identification of contact allergens by animal assay, the guinea pig maximisation test method. *J. Invest. Dermatol.*, **52**, 268–276.

MAISEY, J. and MILLER, K. (1984) Assessment of the ability of mice fed on vitamin A supplemented diet to respond to a variety of contact sensitizers. *Contact Derm.*, **15**, 17–23.

MARZULLI, F. and MAGUIRE, H.C. (1982) Usefulness and limitations of various guinea-pig test methods in detecting human skin sensitizers – validation of guinea-pig tests for skin hypersensitivity. *Fd Chem. Toxicol.*, **20**, 67–74.

MURANAKE, M., SUZUKI, S., KOIZUI, K., IGARASHI, H., OKUMURA, H., TAKEDA, K., *et al.* (1978) Benzylpenicillin preparations can evoke a systemic anaphylactic reaction in guinea pigs. *J. Allergy Clin. Imunol.*, **62**, 276–282.

OVARY, Z. (1951a) Quantitative studies in passive cutaneous anaphylaxis of the guinea-pig. *Int. Archs Allergy*, **3**, 162–174.

OVARY, Z. (1951b) Cutaneous anaphylaxis in the albino rat. *Int. Archs Allergy*, **3**, 293–301.

PAULUHN, J. and EBEN, A. (1991) Validation of a non-invasive technique to assess immediate or delayed onset airway hypersensitivity in guinea-pigs. *J. Appl. Toxicol.*, **11**, 423–431.

SARLO, K. and CLARK, E.D. (1992) A tier approach for evaluating the respiratory allergenicity of low molecular weight chemicals. *Fund. Appl. Toxicol.*, **18**, 107–114.

SARLO, K. and KAROL, M.H. (1994) Guinea pig predictive tests for respiratory allergy. In: *Immunotoxicology and Immunopharmacology*, 2nd edition (Dean, J.H., Luster, M.I., Munson, A.E. and Kimber, I., eds), pp. 703–720. New York: Raven Press.

SCHLEDE, E. and EPPLER, R. (1995) Testing for skin sensitization according to the notification procedure for new chemicals: the Magnusson and Kligman test. *Contact Derm.*, **32**, 1–4.

UFKES, J.G.R., OTTENHOF, M. and AALBERSE, R.C. (1983) A new method for inducing fatal, IgE-mediated, bronchial and cardiovascular anaphylaxis in the rat. *J. Pharmacol. Meth.*, **9**, 175–181.

VERDIER, F., CHAZAL, I. and DESCOTES, J. (1994) Anaphylaxis models in the guinea-pig. *Toxicology*, **93**, 55–61.

VIAL, T. and DESCOTES, J. (1994) Contact sensitization assays in guinea-pigs: are they predictive of the potential for systemic allergic reactions? *Toxicology*, **93**, 63–75.

WATANABE, N. and OVARY, Z. (1977) Antigen and antibody detection by in vivo methods: a reevaluation of passive cutaneous anaphylactic reactions. *J. Immunol. Meth.*, **14**, 381–390.

16

Prediction of Autoimmunity

Autoimmune reactions are relatively rare complications of immunotoxicity, even though growing concern was fuelled by recent epidemics, such as the Spanish toxic oil syndrome or the tryptophan-induced eosinophilia-fasciitis, which were both suggested to be caused by autoimmune mechanisms, and which resulted in many deaths and injured people. A recent survey of adverse reactions induced by medicinal products reported to the French postmarketing drug surveillance (Pharmacovigilance) system found that less than 0.5 per cent of recorded adverse reactions were considered as possible or likely lupus syndromes (Vial *et al.*, 1997b). Autoimmune reactions to occupational chemicals and environmental pollutants seem to be still more rarely recorded, but no reliable data are actually available. A major difficulty when addressing the issue of predicting autoimmune reactions induced by medicinal products and other xenobiotics is the extremely limited understanding of fundamental mechanisms involved in autoimmunity.

Predictability of organ-specific versus systemic autoimmune reactions

As overviewed in Chapter 6 of this volume, it is essential to differentiate autoimmune diseases into organ-specific and systemic diseases, and this also applies to autoimmune reactions of toxic origin.

Organ-specific autoimmune reactions

No models are currently available to predict organ-specific autoimmune reactions routinely induced by xenobiotics. It has generally been very difficult, if not impossible to reproduce in laboratory animals certain drug-induced organ-specific autoimmune reactions that had been previously described in man. For instance, autoimmune haemolytic anaemias to the hypertensive drug α-methyldopa could be inconsistently reproduced in monkeys (Owens *et al.*, 1982), but most attempts in other animal species were unsuccessful. A major limitation to the prediction of organ-specific autoimmune reactions induced by medicinal products and other xenobiotics is the total lack of understanding of the causative mechanism(s).

Systemic autoimmune reactions

Although systemic autoimmune reactions are more frequent, the pathophysiology of these reactions remains ill-known. Various mechanisms have been proposed which served as a basis for the design of potential predictive models.

The possibility that the parent molecule or a metabolite could bind to host proteins, in particular nucleoproteins, was inconclusively suggested to be involved by Whittingham *et al.* (1972). However, there is some evidence that autoimmune reactions, in particular those induced by medicinal products, can actually be auto-allergic, in the sense that reactions develop as a consequence of an immune response directed against tissue, cell or plasma constituents modified by drug binding (Coleman and Sim, 1994). The relevance of findings, such as the immune response directed against captopril-derived antigen in the rat (Foster and Coleman, 1989), remains to be established.

The polyclonal activation of B lymphocytes was shown to account for the experimental autoimmune glomerulonephritis induced by mercuric chloride in the Brown Norway rat and in mice (Kosuda and Bigazzi, 1995), but mercuric chloride has not being conclusively associated with immune-mediated renal injury in humans. Other metals, such as lead, which were shown to induce polyclonal activation of B cells *in vitro* (McCabe and Lawrence, 1990), has never been associated with clinically significant autoimmune reactions, either in animals or in man.

Early studies suggested that a dysregulation of the immune response, for example, an impaired 'suppressor' function of T cells, could be involved, but these often quoted findings with α-methyldopa (Kirtland *et al.*, 1980), could not be reproduced.

Finally, the concept of graft-versus-host like reactions as discussed in Chapter 6 of this volume is particularly interesting as it resulted in the development of the popliteal lymph node assay which is the only foreseeable approach to predict auto-immune reactions induced by medicinal products and other xenobiotics.

The popliteal lymph node assay

Based on similarities between clinical and immunological findings in patients with graft-versus-host (GvH) disease and certain adverse effects induced by certain medicinal products, in particular the anticonvulsant diphenylhydantoin, a GvH-like mechanism was suggested to be involved (Gleichmann, 1982; Gleichmann *et al.*, 1983).

The popliteal lymph node (PLN) assay was derived from the induction of a local GvH reaction in first-generation (F_1) hybrid rats manifesting by an increased volume of the popliteal lymph node after injection of $1 \times 10^6–10^7$ parental splenocytes into the hind footpad (Ford *et al.*, 1970). The injection of the anticonvulsant diphenylhydantoin into the hind footpad of mice induced a similar increase in the popliteal lymph node weight (Gleichmann, 1982) and these findings paved the way for the subsequent development of the PLN assay (Kammüller and Seinen, 1989; Kammüller *et al.*, 1989; Verdier *et al.*, 1990).

Method

The popliteal lymph node assay is performed in young adult mice or rats. No data are available suggesting a difference in responses according to gender or species (namely mice or rats). However, because of its larger size, the rat is often considered more

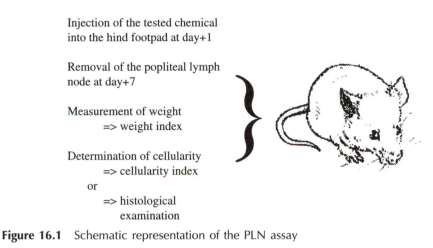

Injection of the tested chemical
into the hind footpad at day+1

Removal of the popliteal lymph
node at day+7

Measurement of weight
=> weight index

Determination of cellularity
=> cellularity index
or
=> histological
examination

Figure 16.1 Schematic representation of the PLN assay

appropriate in order to avoid artifacts related to lymph node removal. Inbred mice, e.g. Balb/c and C57Bl/10 mice, were shown to respond slightly better than outbred mice, but DBA/2 mice were suggested to be poor responders (Kammüller *et al.*, 1989). In contrast, no significant strain differences were observed when comparing four inbred and two outbred rat strains (Patriarca *et al.*, 1994).

The PLN assay is performed in the following several steps:

- *Preparation of the test article solution.* Selection of the vehicle depends on the chemical properties of the tested chemical. Saline and dimethylsulphoxide are the most frequently used vehicles. Saline should however be prefered as DMSO is a primary irritant. Ethanol and acetone induce false positive responses due to their primary irritant effect.

- *Injection of the solution into the right hind footpad.* One mg of the tested chemical in 10 µl of DMSO or 50 µl of saline, is injected to mice, or 5 µg in 50 µl of DMSO or saline, to rats. No data are available suggesting that the injected volume is a critical factor. In contrast, dose was shown to be an important factor: by increasing the dose of procainamide, it was possible to induce a positive PLN response (Roger *et al.*, 1994), which is thought to be due to a metabolite instead of the parent molecule. In addition, systemic toxic effects can be produced, e.g. with mercuric chloride.

- *Animals are killed after a 7–21 day rest period and both popliteal lymph nodes are removed.* The weight index is calculated by comparing the weight of the control versus treated lymph nodes. A weight index above 2 is considered a positive response. It is also possible to calculate the cellularity index from a cellular suspension prepared from each popliteal lymph node. The cellularity index is seemingly a more accurate and sensitive endpoint and a cellularity index above 5 is considered a positive response. Alternatively, weighted lymph nodes can be used for histological examination.

Results and perspectives

Every drug which is known to induce systemic autoimmune reactions in human beings has so far been found to induce a positive PLN response (Descotes *et al.*, 1997). It has also been possible to show that reproducible results can be obtained in several laboratories in a blind fashion (Vial *et al.*, 1997a). Interestingly, pretreatment with either the enzyme inducers β-naphtoflavone or phenobarbital, or with S9 mix, led to a positive response induced by procainamide and isoniazid, which produced a false-negative response when injected to untreated animals (Patriarca *et al.*, 1992). These findings are in agreement with the widely held assumption that metabolites instead of isoniazid or procainamide are involved in systemic autoimmune reactions described in human patients.

Nevertheless, a number of pending questions remain to be answered before the PLN assay can be further recommended for the routine prediction of systemic autoimmune reactions. The first critical question concerns the mechanism involved. Although a number of experimental findings are in full agreement with the initial GvH-like hypothesis, no direct evidence is available. Interestingly, histological examination showed that popliteal lymph nodes of mice and rats treated with chemicals inducing a positive response had a blurred architecture similar to that seen in true local GvH reaction (Brouland *et al.*, 1994; De Bakker *et al.*, 1990). In addition, immunological findings in streptozotocin-treated mice and mice with a true GvH reaction were very similar (Krzystyniak *et al.*, 1992a, 1992b). Another issue is the false positive response induced by primary irritants and sensitisers. The conventional histological examination of popliteal lymph nodes could not help to differentiate primary irritants and sensitisers from 'autoimmunogens' (Brouland *et al.*, 1994).

Additional fundamental research efforts are therefore warranted to gain a better understanding of the ongoing process in the popliteal lymph node response, but the PLN assay can still be expected to be a pivotal assay for screening of immunotoxicants inducing certain autoimmune and/or allergic reactions (Bloksma *et al.*, 1995).

Auto-antibodies in conventional toxicity studies

Auto-antibodies are the hallmark of autoimmune diseases, even though the mere presence of auto-antibodies in the sera of laboratory animals as well as human beings is not sufficient to indicate that an autoimmune disease actually developed.

The spectrum of auto-antibodies is extremely wide (Peter and Shoenfeld, 1996) so that a selection of the most relevant auto-antibodies to be routinely assayed in conventional toxicity studies is essential. In addition, sensitive and specific assays are required. Depending on the auto-antibody under scrutiny, various methods have been proposed, including immunofluorescence, immunoprecipitation, agglutination, radioimmunoassay, ELISA and immunoblot assay, and the selection of the most appropriate method is a critical step (Verdier *et al.*, 1997).

In the context of non-clinical immunotoxicity evaluation, the difficulty is that very few data are available on medicinal drugs and other xenobiotics which induced auto-antibodies in the course of repeated administration studies in laboratory animals. Examples of such chemicals include hydralazine, isoniazid, mercury, gold salts and penicillamine (Balazs and Robinson, 1983; Robinson *et al.*, 1984; Ten Veen and Feltkamp, 1972; Weening *et al.*, 1981). Because these findings are greatly dependent on the strain used to take into account genetic susceptibility, they are not useful within the context of routine non-clinical immunotoxicity evaluation and in fact, when assays for the commonest auto-antibodies in human beings, such as antinuclear, antithyroglobulin, antimitochondrial, antiphospholipid and anti-smooth muscle antibodies, are performed at the end of conventional toxicity studies, results are consistently negative.

Animal models of autoimmune diseases

A number of animal models of human autoimmune diseases have been described (Burkhardt and Kalden, 1997). Presumably, because these models have been genuinely designed for the purpose of fundamental research in immunology and immunopharmacology, most animal models are models of organ-specific autoimmune diseases, which again are seldom described as an adverse consequence of drug treatment or chemical exposure.

Animal models of systemic autoimmune diseases

Several genetically defective mouse strains, such as NZB/NZW, (SWR × SJL)F_1, BXSB and MRL-lpr mice, develop a spontaneous systemic disease which resembles human lupus erythematosus (Theofilopoulos and Dixon, 1985). Polyclonal B cell hyperreactivity is consistently found in these models with hyperglobulinaemia and increased production of antibodies, such as IgG_1 and IgG_2. In addition, abnormal T cell function plays a major role. Depending on the model used, mice develop early clinical and histopathological disorders, including glomerulonephritis, lymphoid hyperplasia, thymic atrophy, and decreasing weight gain, and various serological abnormalities such as increased immunoglobulin levels and auto-antibodies. The disease results in significantly shortened survival.

These models have sometimes been used in an attempt to predict the possible negative impact of medicinal products, such as ethinyloestradiol (Verheul *et al.*, 1995) or chemicals, such as TCDD, metals and silicone (White, 1998) on the development of autoimmune diseases.

Animal models of organ-specific autoimmune diseases

The most extensively studied animal models of organ-specific autoimmune diseases include experimental autoimmune uveoretinitis, experimental allergic encephalomyelitis, collagen II induced arthritis, and insulin-dependent diabetes mellitus.

Experimental autoimmune uveoretinitis

Experimental autoimmune uveoretinitis is induced by the immunisation of several animal strains/species with retina-specific antigens (Faure, 1993). The particular interest of this model is that sequestered antigens are concerned and sequestered antigens were considered as key factors for the development of organ-specific autoimmunity in the past.

Experimental allergic encephalomyelitis

Experimental allergic encephalomyelitis is induced by the immunisation of mice, rats or guinea-pigs with myelin antigens, such as myelin basic protein (MBP) and MBP-derived peptides. There is a tight genetic control of susceptibility.

The disease is mainly characterised by ascending paralysis of the hind limbs and histological anomalies including perivascular infiltration of mononuclear cells and areas of acute or chronic demyelination in the spinal cord and the brain. Depending on the experimental protocol (in particular the antigen) and the strain, two forms of the disease have been identified, an acutely progressing and a chronic disease. Although experimental allergic encephalomyelitis is considered to be the best animal model for human multiple sclerosis and is widely used for the early evaluation of new therapeutic modalities (Swanborg, 1995), there is no evidence that it can be a useful model for evaluation of the impact of medicinal products and other xenobiotics on the course of autoimmune diseases.

Collagen II induced arthritis

Collagen II induced arthritis is an animal model for rheumatoid arthritis which is the most common organ-specific autoimmune disease in human beings (Billingham, 1995). This model is induced in mice and rats by the injection of collagen II emulsified in complete Freund's adjuvant. After three weeks, the animals develop joint swelling and oedema.

Insulin-dependent diabetes mellitus

Insulin-dependent diabetes mellitus is a complex situation involving autoreactive B and T cells. The non-obese diabetic (NOD) mouse and the BB rat develop a spontaneous disease similar to insulin-dependent diabetes mellitus. The repeated administration of low-dose streptozotocin was reported to induce diabetes in NOD mice (Like and Rossini, 1976).

Overall, animal models of autoimmune diseases have very seldom been used in the context of immunotoxicology. Specific models using certain chemicals, such as mercuric chloride-induced autoimmune glomerulonephritis in Brown Norway rats (Pelletier *et al.*, 1987) proved useful tools to dissect the fundamental regulatory and effector mechanisms in autoimmunity, but did not provide utilisable information for the purpose of immuno-toxicity evaluation.

References

BALAZS, T. and ROBINSON, C. (1983) Procainamide-induced antinuclear antibodies in beagle dogs. *Toxicol. Appl. Pharmacol.*, **71**, 299–302.

BILLINGHAM, M.E.J. (1995) Adjuvant arthritis: the first model. In: *Mechanisms and Models of Rheumatoid Arthritis* (Henderson, B., Edwards, J.C.W. and Pettipher, E.R., eds), pp. 389–409. London: Associated Press.

BLOKSMA, N., KUBICKA-MURANYI, M., SCHUPPE, H.C., GLEICHMANN, E. and GLEICHMANN, H. (1995) Predictive immunotoxicological test systems: suitability of the popliteal lymph node assay in rats and mice. *Crit. Rev. Toxicol.*, **25**, 369–396.

BROULAND, J.P., VERDIER, F., PATRIARCA, C., VIAL, T. and DESCOTES, J. (1994) Morphology of popliteal lymph node responses in Brown-Norway rats. *J. Toxicol. Environ. Health*, **41**, 95–108.

BURKHARDT, H. and KALDEN, J.R. (1997) Animal models of autoimmune diseases. *Rheumatol. Int.*, **17**, 91–99.

COLEMAN, J.W. and SIM, E. (1994) Autoallergic responses to drugs. Mechanistic aspects. In: *Imunotoxicology and Immunopharmacology*, 2nd edition (Dean, J.H., Luster, M.I., Munson, A.E. and Kimber, I., eds), pp. 533–572. New York: Raven Press.

DE BAKKER, J.M., KAMMÜLLER, M.E., MULLER, E.S.M., LAM, S.W., SEINEN, W. and BLOKSMA, N. (1990) Kinetics and morphology of chemically induced popliteal lymph node reactions compared with antigen-, mitogen-, and graft-versus-host reaction-induced responses. *Virchows Arch. B Cell. Pathol.*, **58**, 279–287.

DESCOTES, J., PATRIARCA, C., VIAL, T. and VERDIER, F. (1997) The popliteal lymph node assay in 1996. *Toxicology*, **119**, 45–49.

FAURE, J.P. (1993) Autoimmune diseases of the retina. In: *The Molecular Pathology of Auto-immune Diseases* (Bona, C.A., Siminovitch, K.A., Zanetti, M. and Theofilopoulos, A.N., eds), pp. 651–672. Chur: Harwood.

FORD, W.L., BURR, W. and SIMONSEN, M. (1970) A lymph node weight assay for graft-versus-host activity of rat lymphoid cells. *Transplantation*, **10**, 258–266.

FOSTER, A.L. and COLEMAN, J.W. (1989) A rat model of captopril immunogenicity. *Clin. Exp. Immunol.*, **75**, 161–175.

GLEICHMANN, H. (1982) Studies on the mechanism of drug sensitization: T-cell-dependent popliteal lymph node reaction to diphenylhydantoin. *Clin. Immunol. Immunopathol.*, **18**, 203–211.

GLEICHMANN, H., PALS, S.T. and RADASZKIEWICZ, T. (1983) T-cell-dependent B-cell prolif-eration and activation induced by administration of the drug diphenylhydantoin to mice. *Hematol. Oncol.*, **1**, 165–176.

GOTER ROBISON, C.J., BALACZ, T. and EGOROV, I.K. (1986) Mercuric chloride-, gold sodium thiomalate- and D-penicillamine-induced antinuclear antibodies in mice. *Toxicol. Appl. Pharmacol.*, **86**, 159–169.

HANG, L., AGUADO, M.Y., DIXON, F.J. and THEOFILOPOULOS, A.N. (1985) Induction of severe autoimmune disease in normal mice by simultaneous action of multiple immuno-stimulators. *J. Exp. Med.*, **161**, 423–428.

KAMMÜLLER, M.E. and SEINEN, W. (1989) Structural requirements for hydantoin and 2-thiohydantoins to induce lymphoproliferative popliteal lymph node reations in the mouse. *Int. J. Immunopharmac.*, **10**, 997–1010.

KAMMÜLLER, M.E., THOMAS, C., DE BAKKER, J.M., BLOKSMA, N. and SEINEN, W. (1989) The popliteal lymph node assay in mice to screen for the immune disregulating potential of chemicals – A preliminary study. *Int. J. Immunopharmac.*, **11**, 293–300.

KIRTLAND, H.H., MOHLER, D.N. and HORWITZ, D.A. (1980) Methyldopa inhibition of suppressor-lymphocyte function. A proposed cause of autoimmune hemolytic anemia. *N. Engl. J. Med.*, **302**, 825–832.

KOSUDA, L.L. and BIGAZZI, P.E. (1995) Chemical-induced autoimmunity. In: *Experimental Immunotoxicology* (Smialowicz, R.J. and Holsapple, M.P., eds), pp. 419–468. Boca Raton: CRC Press.

KRZYSTYNIAK, K., BROULAND, J., PANAYE, G., PATRIARCA, V., VERDIER, F., DESCOTES, J., and REVILLARD, J-P. (1992a) Changes in CD4$^+$ and CD8$^+$ lymphocyte subsets by streptozotocin in murine popliteal lymph node (PLN) test. *J. Autoimmun.*, **5**, 183–197.

KRZYSTYNIAK, K., PANAYE, G., DESCOTES, J., and REVILLARD, J-P. (1992b) Changes in lymphocyte subsets during acute local graft-versus-host reaction in H-2 incompatible murine F1 hybrids. *Immunopharmacol. Immunotoxicol.*, **14**, 865–882.

LIKE, A.A. and ROSSINI, A.A. (1976) Strepozotocin-induced pancreatic insulitis: new model of diabetes mellitus. *Science*, **193**, 415–417.

MCCABE, M.J. and LAWRENCE, D.A. (1990) The heavy metal lead exhibits B cell-stimulatory factor activity by enhancing Ia expression and differentiation. *J. Immunol.*, **145**, 671–677.

OWENS, N.A., HUI, H.L. and GREEN, F.A. (1982) Induction of direct Coombs positivity with alpha-methyldopa in chimpanzees. *J. Med.*, **13**, 472–477.

PATRIARCA, C., VERDIER, F., BROULAND, J.P. and DESCOTES, J. (1992) Popliteal lymph node response to procainamide and isoniazid. Role of β-naphtoflavone, phenobarbitone and S-9 mix pretreatment. *Toxicol. Letters*, **66**, 21–28.

PATRIARCA, C., VERDIER, V., BROULAND, J.P., VIAL, T. and DESCOTES, J. (1994) Compari-son of popliteal lymph node responses in various strains of rats. *Hum. Exp. Toxicol.*, **13**, 455–460.

PELLETIER, L., HIRSCH, F., ROSSER, J., DRUET, E. and DRUET, P. (1987) Experimental mercury-induced glomerulonephritis. *Springer Semin. Immunopathol.*, **9**, 359–369.

PETER, J.B. and SHOENFELD, Y. (1996) *Autoantibodies*. Amsterdam: Elsevier.

ROBINSON, C.J.G., ABRAHAM, A.A. and BALAZS, T. (1984) Induction of anti-nuclear antibodies by mercuric chloride in mice. *Exp. Clin. Immunol.*, **58**, 300–306.

ROGER, I., DOUVIN, D., BÉCOURT, N. and LEGRAIN, B. (1994) Procainamide (PA) and popliteal lymph node assay (PLNA): false negative response? *Toxicologist*, **14**, 324.

SWANBORG, R.H. (1995) Animal models of human disease: experimental autoimmune encephalomyelitis in rodents as a model for human demyelinating disease. *Clin. Immunol. Immunopathol.*, **77**, 4–13.

TEN VEEN, J.H. and FELTKAMP, T.E.W. (1972) Studies on drug induced lupus erythematosus in mice. I. Drug induced antinuclear antibodies (ANA). *Clin. Exp. Immunol.*, **11**, 265–276.

THEOFILOPOULOS, A.N. and DIXON, F.J. (1985) Murine models of systemic lupus erythematosus. *Adv. Immunol.*, **35**, 269–290.

VERDIER, F., VIRAT, M. and DESCOTES, J. (1990) Applicability of the popliteal lymph node assay in the Brown-Norway rat. *Immunopharmacol. Immunotoxicol.*, **12**, 669–677.

VERDIER, F., PATRIARCA, C. and DESCOTES, J. (1997) Autoantibodies in conventional toxicity testing. *Toxicology*, **119**, 51–58.

VERHEUL, H.A., VERVELD, M., HOEFAKKER, S. and SCHUURS, A.H. (1995) Effects of ethinyloestradiol on the course of spontaneous autoimmune disease in NZB/W and NOD mice. *Immunopharmacol. Immunotoxicol.*, **17**, 163–180.

VIAL, T., LEGRAIN, B., CARLEER, J., VERDIER, F. and DESCOTES, J. (1997a) The popliteal lymph node assay: results of a preliminary interlaboratory validation. *Toxicology*, **122**, 213–218.

VIAL, T., NICOLAS, B. and DESCOTES, J. (1997b) Drug-induced autoimmunity. Experience of the French Pharmacovigilance system. *Toxicology*, **119**, 23–27.

WEENING, G.G., HOEDEMAEKER, P.J. and BAKKER, W.W. (1981) Immunoregulation and antinuclear antibodies in mercury induced glomerulopathy in the rat. *Clin. Exp. Immunol.*, **45**, 64–71.

WHITE, K.L. (1998) Use of Brown Norway rats and NZBxW mouse models of systemic lupus erythematosus to assess effects of silicone gel, metals, and other xenobiotics on autoimmune disease. *Toxicol. Sci.*, **42**, suppl. 1S, 403.

WHITTINGHAM, S., MACKAY, I.R., WHITEWORTH, J.A. and SLOMAN, G. (1972) Antinuclear antibody response to procainamide in man and laboratory animals. *Am. Heart J.*, **84**, 228–234.

PART III

Trends and Perspectives in Immunotoxicology

17

Immunotoxicity Risk Assessment

As already mentioned in this volume, there is an enormous body of evidence that medicinal products, and occupational as well as environmental chemicals, can exert immunotoxic effects, either the suppression/depression or the stimulation of the immune responses of treated/exposed hosts, or the induction of hypersensitivity (allergic) and autoimmunity reactions of varying frequency and severity. As nearly all of these data were obtained in laboratory animals, a major limitation to immunotoxicity risk assesment is the current lack of adequate human data.

Initially, immunotoxicologists as toxicologists, focused their efforts on the identification of hazard, namely immunotoxic effects, and no attention was paid to immunotoxicity risk assessment. Therefore, most immunotoxicity data from animal studies are at best of limited relevance because they were obtained in experimental conditions remote from those of human exposure. In addition, until recently immunotoxicity was governed by immunological considerations much more than toxicological considerations. Fortunately, immunotoxicologists increasingly turned to assessing the risk in relation to immunotoxic effects (ILSI, 1995; Luster *et al.*, 1994b; Selgrade *et al.*, 1995; Trizio *et al.*, 1988) and risk assessment is now emerging as a new and important area of immunotoxicology.

This short chapter is an attempt to overview the current status and trends in the risk assessment for immunotoxicology.

Toxicity risk assessment

Toxicity risk assessment is the process of evaluating the nature and likelihood of adverse effects that may occur following exposure to a toxicant (Scala, 1991). Although it is very common for people to use the terms 'hazard' and 'risk' interchangeably, their definition is clearly different.

Hazard (an equivalent of toxicity for chemical substances) refers to the potential adverse effects of a certain chemical. Based on its intrinsic physical, chemical and biochemical properties, a chemical can exert hazardous properties. Risk is therefore the likelihood for these hazardous properties to result in toxic effects in an individual or a population in given conditions of exposure, such as the duration and magnitude of the exposure. In contrast to hazard which can be objectively determined provided that adequate methodologies are selected and used, uncertainty is a major and unavoidable feature of risk.

Risk assessment is a largely scientific process including four components, as initially defined by the US National Research Council (1983).

Identification of hazard

The identification of hazard is the very first component of risk assessment. In this step, adverse effects of the chemical are determined, including mortality, reproductive and developmental toxicity, neurotoxicity, carcinogenicity, and very inconsistently immunotoxicity. The data are usually derived from the results of animal studies, but also from human findings in epidemiological studies or clinical surveys, as well as *in vitro* experiments.

Dose–response assessment

In this step, the quantitative nature of the relationship between the dose and the biological (toxic) response is considered. Difficulties are the extrapolation from animals to human beings and from high to low doses. In fact, it is often impossible to estimate the dose–response directly from human data. Major issues are related to the identification of a threshold and appropriate simulation models.

Exposure

Exposure, namely the duration, magnitude, frequency and routes of exposures as well as the number and characteristics of exposed subjects (at risk subjects, age, gender, occupation, etc.), is usually the most neglected aspect of the risk assessment process.

Risk characterisation

This is the final step of the process during which the incidence of adverse effects is estimated in a given population. The reliability and acceptability of the estimate depend on the quality of the whole process and on the identification of uncertainties.

This is fundamentally a science-based judgement, but since the final results of the risk assessment process are aimed at determining adequate measures to control or remove the risk (risk management), the perception of risk should also be taken into consideration. The term 'risk analysis' is often used to include risk assessment and the perception of risk.

Approaches in immunotoxicity risk assessment

Immunotoxicity risk assessment should typically include the four components of the risk assessment process. The identification of hazard ('immunotoxicity') has been the focus

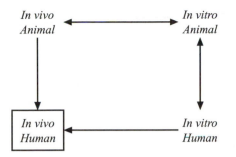

Figure 17.1 The parallelogram approach for immunotoxicity risk assessment

of many studies in the past two decades. Due to extensive efforts to standardise and validate animal models, and evaluate the relevance of results for the purpose of hazard identification, a tiered battery of tests has been proposed and is widely accepted (Luster *et al.*, 1992, 1994a). Because the dose-response relationship of most immunotoxic effects is ill-understood or atypical ('bell-shaped curve'), much remains to be done for designing appropriate quantitative experimental models useful for immunotoxicity risk assessment. Exposure data are not specific to immunotoxicity risk assessment.

Finally, the key issue remains immunotoxicity risk characterisation. Unfortunately, the near total lack of human data is a major difficulty. The parallelogram approach was proposed to overcome this difficulty. It is possible to compare the results of *in vitro* and *in vivo* animal studies using the same or similar endpoints of immunotoxicity, as it is possible to compare the results of *in vitro* studies in animals and in man. Based on the combined comparison of these results, an improved extrapolation from animals to humans is supposedly achievable.

The parallelogram approach has been used for the immunotoxicity risk assessment of a limited number of immunotoxicants, such as ozone and dioxin. The most illustrative example is however the immunotoxicity risk assessment of UVB exposure (Van Loveren *et al.*, 1997). The hazard of UVB in relation to immunotoxicity is clearly established from the results of many *in vivo* and *in vitro* studies. The relation between the dose (UVB irradiation) and the biological response (resistance to *Listeria monocytogenes* experimental infection) was determined. Based on the results of artificial exposure to UVB, it was possible to calculate the risk for impaired resistance to infections following a given duration of human UVB exposure.

References

ILSI (1995) Immunotoxicity testing and risk assessment: summary of a workshop. *Fd Chem. Toxicol.* **33**, 887–894.

LUSTER, M.I., PORTIER, C., PAIT, D.G., WHITE, K.L., GENNINGS, C., MUNSON, A.E. and ROSENTHAL, G.J. (1992) Risk assessment in immunotoxicology. I. Sensitivity and predictability of immune tests. *Fund. Appl. Toxicol.*, **18**, 200–210.

LUSTER, M.I., PORTIER, C., PAIT, D.G., ROSENTHAL, G.J., GERMOLEC, D.R., CORSINI, E., *et al.* (1994a) Risk assessment in immunotoxicology. II. Relationships between immune and host resistance tests. *Fund. Appl. Toxicol.*, **21**, 71–82.

LUSTER, M.I., SELGRADE, M.K., GERMOLEC, D.R., BURLESON, F.G., KAYAMA, F., COMMENT, C.E. and WILMER, J.L. (1994b) Experimental studies on immunosuppression. Approaches and application in risk assessment. In: *Immunotoxicology and Immunopharmacology*,

2nd edition (Dean, J.H., Luster, M.I., Munson, A.E. and Kimber, I., eds) pp. 51–69. New York: Raven Press.

SCALA, R.A. (1991) Risk assessment. In: *Casarrett and Doull's Toxicology – The Basic Science of Poisons*, 5th edition (Amdur, R.O., Doull, J. and Klaassen, C.D., eds), pp. 985–996. New York: Pergamon Press.

SELGRADE, M.J.K., COOPER, K.D., DEVLIN, R.B., VAN LOVEREN, H., BIAGINI, R.E. and LUSTER, M.I. (1995) Immunotoxicity – bridging the gap between animal research and human health effects. *Fund. Appl. Toxicol.*, **24**, 13–21.

TRIZIO, D., BASKETTER, D.A., BOTHAM, P.A., GRAEPEL, P.H., LAMBRE, C., MAGDA, S.J., *et al.* (1988) Identification of immunotoxic effects of chemicals and assessment of their relevance to man. *Food Chem. Toxicol.*, **26**, 527–539.

US NATIONAL RESEARCH COUNCIL (1983) *Risk Assessment in the Federal Government: Managing the Process.* Washington DC: National Academy Press.

VAN LOVEREN, H., GOETTSCH, W., SLOB, W. and GARSSEN, J. (1997) Risk assessment for the harmful effects of immunotoxic agents on the immunological resistance to infectious diseases: ultraviolet light as an example. *Toxicology*, **119**, 59–64.

18

New Methods in Immunotoxicology

Because immunotoxicology is a new area of toxicology, immunotoxicologists have no predictive tools to address various and critical aspects of immunotoxicity, despite a lot of past and present research efforts. New methods, in particular *in vitro* assays and genetically modified animals, are expected to improve the validity of current evaluation modalities. However, these new methods have essentially been used in the context of fundamental research and mechanistic immunotoxicity studies. No *in vitro* assay or genetically-modified animal model have so far been adequately validated for use in the routine non-clinical immunotoxicity evaluation of medicinal products and other xenobiotics. However, extremely rapid progress in this area should logically result in the availability of totally new immunotoxicology methods.

In vitro methods for immunotoxicity evaluation

A number of assays in current use for non-clinical as well as human immunotoxicology studies are in fact *ex-vivo* assays, that is to say assays which are performed with cells from animals or human beings after *in vivo* treatment. *Ex-vivo* assays can therefore easily be performed totally *in vitro*, as there are no or minimal technical differences whether the same assay is performed *ex-vivo* or *in vitro*.

Compared to *in vivo* assays, *in vitro* assays have several advantages and limitations. Major advantages are the possibilities to use a wider range of concentrations, to duplicate/triplicate the same assay, to test more easily chemicals which pose certain problems of safety (e.g. carcinogenic or genotoxic chemicals), and also to include highly sophisticated endpoints. On the other hand, major limitations of *in vitro* assays are the lack of biotransformation in test tubes or cultured cells, the possible non-specific cytopathic/toxic effects of tested chemicals resulting in misleading findings, and the lack of external influences (such as neuro-endocrine interactions) on the biological response.

In vitro assays have been suggested to be useful to improve the integration of immunotoxicology in drug discovery and development (Dean *et al.*, 1994) and this could probably apply to any type of novel chemical entity as well, whatever the potential utilisation. In this context, expected benefits from *in vitro* immunotoxicology include the quick confirmation and understanding of findings made in conventional *in vivo* toxicity animal testing and the availability of a rapid screen for the improved safety evaluation of new chemical entities.

Even though much remains to be done in this field, a number of *in vitro* possibilities have been identified (Sundwall *et al.*, 1994). They include immune function tests, which are mostly derived from *ex-vivo* tests, and models of contact sensitivity and molecular immunotoxicology testing, which are in fact both essentially research tools at the present time.

In vitro *immune function tests*

Several *ex-vivo* immune function tests have been used *in vitro* for immunotoxicity evaluation. The most commonly used tests include mitogen-induced lymphocyte proliferation, *in vitro* antibody producing systems similar to the plaque-forming cell assay, and NK cell activity (Cornacoff *et al.*, 1989; Lang *et al.*, 1993; Lebrec *et al.*, 1995; Wood *et al.*, 1992). The methods are in no way different from those recommended when performing the assays *ex-vivo*.

Interestingly, these *in vitro* tests were suggested to be useful even for chemicals which require prior activation to induce their specific biological responses. This can be achieved, for instance, by the use of S9 mix preincubation (Archer, 1982; Fautz and Miltenburger, 1993; Tucker *et al.*, 1982).

Although comparisons of *ex-vivo* and *in vitro* results obtained with the same chemical or group of chemicals were found satisfactory by a few investigators (Lang *et al.*, 1993; Lebrec *et al.*, 1995), very few studies compared a reasonably large series of chemicals (Archer *et al.*, 1978; Fautz and Miltenburger, 1993), so that the predictive value of *in vitro* immune function tests for non-clinical immunotoxicity evaluation remains to be fully established.

In vitro *models of contact hypersensitivity*

During the past two decades, studies on the sensitising potential of xenobiotics have been entirely based on *in vivo* animal testing, especially guinea-pig contact sensitisation assays. No *in vitro* tests predictive of contact sensitisation have so far been validated despite various attempts to design such tools (Elmets, 1996; Parish, 1986).

To date, avenues of research focused on the cytokine response of cultured keratinocytes induced by haptens, on alterations in adhesion molecules on keratinocytes and epidermal Langerhans cells, and on hapten-mediated T cell activation by Langerhans cells. Another interesting approach is the search for structure–activity relationships (Barratt *et al.*, 1994).

In fact, lymphocyte proliferation (lymphocyte blastogenesis assay) is the only assay which has been used in the context of non-clinical immunotoxicity evaluation. This is an *ex-vivo* assay (Kashima *et al.*, 1993) and its predictive value as a totally *in vivo* assay is not established.

In vitro *molecular immunotoxicology testing*

The extraordinary advances in molecular biology techniques led to the recommendation of the introduction of molecular endpoints in immunotoxicology (Vandebriel *et al.*, 1996).

Besides genetically modified animals as discussed later in this chapter, molecular immuno-toxicology testing essentially focused on the mRNA expression of cytokines. However, considerable research effort needs to be paid to the refinement, standardisation and valida-tion of these tests before they can come out from the area of fundamental research as suggested by the most recent findings (Vandebriel *et al.*, 1998).

New animal models

Recent dramatic progress in the development of new animal models for use in various areas of biology, including immunology and toxicology, opened interesting avenues of research. These new animal models which could prove useful in immunotoxicology, include SCID mice, and genetically modified (knock-out and transgenic) animals (Løvik, 1997).

SCID mice

Severe combined immune deficiency (SCID) is a congenital disease syndrome first recog-nised in human infants in the mid-1950s. SCID mice were first described in CB-17 mice which are an immunoglobulin allotype variant of Balb/c mice (Bosma and Carroll, 1991). Because of their defect in antigen receptor gene rearrangement, mice have no functional B and T lymphocytes. They have profound lymphopenia and the size of their thymus is only 1–10 per cent of the normal thymus. SCID mice are unable to mount an immune response to T-dependent and T-independent antigens, but NK cell activity and the bone marrow are normal. Due to their immune defect, SCID mice require specific care: they must be kept in a protected environment to avoid contact with microbial pathogens.

A major interest in the use of SCID mice for (immuno-)toxicological investigations is the possibility to graft human tumours and liver, thyroid or skin tissues as well as human immunocompetent cells. SCID reconstituted with lymphoid tissue, peripheral blood mono-nuclear cells, and human foetal tissues can mount a normal or nearly normal immune response.

Despite some limitations and difficulties still to overcome, SCID mice are expected to become valuable models for immunotoxicology evaluation (Løvik, 1995). The effects of dioxin and cyclosporin A (Pollock *et al.*, 1994), or 2-acetyl-4(5)-tetrahydroxybutyl-imidazole and di-n-butyltin dichloride (De Heer *et al.*, 1995) have been studied in SCID mice.

However, more research is warranted to establish the value of this model for immuno-toxicity evaluation, in particular to establish whether the engraftment of human lymphoid cells is instrumental for an improved extrapolation from animals to man, as previously suggested (Van Loveren and De Heer, 1995).

Genetically modified animals

A major achievement in biological research is the development of animals with new functional genes added to their genome, and animals with altered or inactivated genes (transgenic, knock-out and knock-in animals). The variety of genetically modified ani-mals is growing daily (Shasty, 1995).

In the field of toxicology, there is particular interest in genetically modified animals to increase their sensitivity to screen for toxic effects, such as mutagenesis (Mirsalis *et al.*,

1994) or to investigate mechanisms, such as Ah-receptor knock-out mice (Fernandez-Salguero *et al.*, 1995). A number of cytokine knock-out mice have been developed. Their use in mechanistic studies is being investigated in order to show their superiority over conventional animals for immunotoxicity evaluation (Ryffel, 1997).

Perspectives

As the trend in the media and the public is in favour of *in vitro* assays for toxicological evaluation and despite the intrinsic difficulties in performing such assays, *in vitro* immunotoxicity assays are likely to be more and more popular in the future. At the present time, *in vitro* assays are potentially useful only as a screen for closely related chemicals and for the investigation of mechanisms. Much remains to be done before *in vitro* assays can prove useful in the routine safety assessment of new chemical entities and it is logical to believe they may not provide sufficient information to avoid completely the use of conventional *in vivo* animal toxicity testing. A critical, although seldom addressed issue, is the regulatory acceptability of new toxicity methods, even though this issue is obviously not restricted to new immunotoxicity methods (Dean, 1997).

References

ARCHER, D.L. (1982) New approaches to immunotoxicity testing. *Environ. Health Perspect.*, **43**, 109–113.

ARCHER, D.L., SMITH, B.G. and BUKOVIC-WESS, J.A. (1978) Use of an vitro antibody producing system for recognizing potentially immunosuppressive compounds. *Int. Arch. Allergy Appl. Immunol.*, **56**, 90–93.

BARRATT, M.D., BASKETTER, D.A., CHAMBERLAIN, M., ADMANS, G. and LANGOWSKI, J. (1994) An expert system rule base for identifying contact allergens. *Toxicol. in vitro*, **8**, 1053–1060.

BOSMA, M.J. and CARROLL, A.M. (1991) The SCID mouse mutant: definition, characterization, and potential uses. *Annu. Rev. Immunol.*, **9**, 323–350.

CORNACOFF, J.B., TUCKER, A.N. and DEAN, J.H. (1989) Development of a human peripheral blood mononuclear leukocyte cell model for immunotoxicity evaluation. *Toxicl. in vitro*, **2**, 81–90.

DEAN, J.H. (1997) Issues with introducing new immunotoxicology methods into the safety assessment of pharmaceuticals. *Toxicology*, **119**, 95–101.

DEAN, J.H., CORNACOFF, J.B., HALEY, P.J. and HINKS, J.R. (1994) The integration of immunotoxicology in drug discovery and development: investigative and *in vitro* possibilities. *Toxic. in vitro*, **8**, 939–944.

DE HEER, C., SCHUURMAN, H.J., HOUBEN, G.F., PIETERS, R.H.H., PENNINKS, A.H. and VAN LOVEREN, H. (1995) The SCID-hu mouse as a tool in immunotoxicological risk assessment: effects of 2-acetyl-4(5)-tetrahydroxybutyl-imidazole (THI) and di-n-butyltin dichloride (DBTC) on the human thymus in SCID-hu mice. *Toxicology*, **100**, 203–211.

ELMETS, C.A. (1996) Progress in the development of an *in vitro* assay for contact allergens: results of the Avon program project experience. *Toxicol. in vitro*, **9**, 223–235.

FAUTZ, R. and MILTENBURGER, H.G. (1993) Immunotoxicity screening *in vitro* using an economical multiple endpoint approach. *Toxic. in vitro*, **7**, 305–310.

FERNANDEZ-SALGUERO, P., PINEAU, T., HILBERT, D.M., *et al.* (1995) Immune system impairment and hepatic fibrosis in mice lacking the dioxin-binding Ah-receptor. *Science*, **268**, 722–726.

KASHIMA, R., OKADA, J., IEDA, Y. and YOSHIZUKA, N. (1993) Challenge assay *in vitro* using lymphocyte blastogenesis for the contact hypersensitivity assay. *Fd Chem. Toxicol.*, **31**, 756–766.

LANG, D., MEIER, K.L. and LUSTER, M.I. (1993) Comparative effects of immunotoxic chemicals on *in vitro* proliferative responses of human and rodent lymphocytes. *Fund. Appl. Toxicol.*, **21**, 535–545.

LEBREC, H., ROGER, R., BLOT, C., BURLESON, G.R., BOHUON, C. and PALLARDY, M. (1995) Immunotoxicological investigation using pharmaceutical drugs. In vitro evaluation of immune effects using rodent or human immune cells. *Toxicology*, **96**, 147–156.

LØVIK, M. (1995) The SCID (severe immunodeficiency) mouse – its biology and use in immunotoxicological research. *Arch. Toxicol.*, **Suppl. 17**, 455–467.

LØVIK, M. (1997) Mutant and transgenic mice in immunotoxicology: an introduction. *Toxicology*, **119**, 65–76.

MIRSALIS, J.C., MONFORTE, J.A. and WNEGAR, R.A. (1994) Transgenic animal models for measuring mutations *in vivo*. *Crit. Rev. Toxicol.*, **24**, 255–280.

PARISH, W.E. (1986) Evaluation of *in vitro* predictive tests for irritation and allergic sensitization. *Fd Chem. Toxicol.*, **24**, 481–494.

POLLOCK, P.L., GERMOLEC, D.R., COMMENT, C.E., ROSENTHAL, G.J. and LUSTER, M.I. (1994) Development of human lymphocyte-engrafted SCID mice as a model for immunotoxicity assessment. *Fund. Appl. Toxicol.*, **22**, 130–138.

RYFFEL, B. (1997) Impact of knockout mice in toxicology. *CRC Crit. Rev. Toxicol.*, **27**, 135–154.

SHASTY, B.S. (1995) Genetic knockouts in mice: an update. *Experientia*, **51**, 1028–1039.

SUNDWALL, A., ANDERSSON, B., BALLS, M., *et al.* (1994) Immunotoxicology and *in vitro* possibilities. *Toxic. in Vitro*, **8**, 1067–1074.

TUCKER, A.N., SANDERS, V.M., HALLETT, P., KAUFFMANN, B.M. and MUNSON, A.E. (1982) *In vitro* immunotoxicological assays for detection of compounds requiring metabolic activation. *Environ. Health Perspect.*, **43**, 123–127.

VANDEBRIEL, R.J., MEREDITH, C., SCOTT, M.P., GLEICHMANN, E., BLOKSMA, N., VANT'ERVE E.H.M., *et al.* (1996) Early indicators of immunotoxicity: development of molecular biological test batteries. *Hum. Exp. Toxicol.*, **15**, Suppl. 1, 2–9.

VANDEBRIEL, R.J., VAN LOVEREN, H. and MEREDITH, C. (1998) Altered cytokine (receptor) mRNA expression as a tool in immunotoxicology. Toxicology, in press.

VAN LOVEREN, H. and DE HEER, C. (1995) The SCID mouse as a tool to bridge the gap between animal and human responses. *Arch. Toxicol.*, **Suppl. 17**, 468–471.

WOOD, S.C., KARRAS, J.G. and HOLSAPPLE, M.J. (1992) Integration of the human lymphocyte into immunotoxicological investigations. *Fund. Appl. Toxicol.*, **18**, 450–459.

19

Human Immunotoxicology

Immunotoxicology was and still is primarily an area of experimental toxicology. There-fore, immunotoxicologists focused their efforts on unravelling the fundamental mechan-isms of immunotoxicity and designing new methods and strategies to conduct relevant non-clinical immunotoxicity evaluations of xenobiotics. In recent years however, the clinical aspects of immunotoxicity have become a matter of growing interest and invest-igation (Salvaggio, 1990; Burrel *et al.*, 1992; Newcombe *et al.*, 1992).

At least three main objectives fall into the scope of human immunotoxicology: the identification of immunotoxicants in human beings; the surveillance of selected groups of the population exposed to recognised immunotoxicants; and the validation of current non-clinical immunotoxicity assays and animal models. To attain these objectives, a major issue is the still unmet need for better, more selective and sensitive biomarkers of immunotoxicity to be used in the clinical trials of new medicinal products as well as environmental epidemiology studies.

Identification of immunotoxicants in human beings

Identifying chemical immunotoxicants in human beings is certainly the most important and timely area of human immunotoxicology. In sharp contrast to the wealth of animal data, very limited information is available on the immunological changes induced by the treatment of patients with medicinal products, and by inadvertent exposure to occupa-tional chemicals and environmental pollutants. In addition, still less is known of the adverse health consequences (diseases) actually associated with such immunological changes in realistic conditions of treatment or exposure.

Environmental epidemiology studies

Ideally, the identification of immunotoxicants in human beings should be based on epide-miological studies. Environmental epidemiology can be defined as 'the study of environ-mental factors that influence the distribution and determinants of disease in human populations' (Griffith and Aldrich, 1993). Clearly, immunotoxicants are environmental factors which can potentially result in human diseases.

Experimental versus observational epidemiological methods

Various types of methods have been designed to find associations between exposure and disease: they are divided into experimental and observational methods (Griffith *et al.*, 1993; Hernberg, 1992).

The experimental methods include clinical trials which use patients as subjects (in particular for evaluation of the efficacy of novel medicinal products), field trials which use healthy patients in a controlled environmental situation (for instance to evaluate the preventive value of a new modality of nutritional supplementation in a fraction of the population), and community trials (for instance to evaluate either the efficacy or safety of new processes, such as the fluorination of drinking water). These methods are also often called intervention studies in the sense that the investigators typically change the conditions of exposure.

Observational methods include descriptive, analytical and cross-sectional studies. Descriptive studies are confined to describing a population of interest with regard to potential health problems. Because they are unable to establish causality between exposure and disease, these often called 'fishing expeditions' may identify purely random, statistically significant events leading to erroneous or misleading conclusions. Analytical studies are the core epidemiological studies, including longitudinal (cohort and case-control) studies and cross-sectional studies. When properly designed, these methods are critical for establishing causality between exposure and disease.

Use of epidemiological studies in immunotoxicology

Despite enormous methodological efforts and scientific achievements in the past decades, epidemiology has its own limits (Taubes, 1995). With regard to immunotoxicological issues, a number of difficulties must be overcome prior to conducting epidemiological studies (Biagini, 1998; Biagini *et al.*, 1994):

- Most available clinical immunology assays have neither been extensively standardised, as for instance clinical chemistry assays have been in the past, nor adequately validated for the benefit of immunotoxicity risk assessment.

- The so-called 'functional reserve capacity' of the human immune system is unsizable so that it remains to be established whether recorded changes in a given endpoint, such as a 20 per cent decrease in the level of antibody response against a specific antigen, are truly indicative of immunotoxicity, or merely theoretical and speculative.

- The appropriate selection of controls is absolutely essential to avoid factors which interfere with immune competence, such as age (e.g. infants and elderly), gender, smoking, nutritional status (e.g. malnutrition and vitamin deficiencies), or underlying illnesses (e.g. HIV infection, cancer or asthma). As so many confounding factors have to be taken into account, the size of both the exposed and the control populations in epidemiological immunotoxicity studies have often to be much larger than in other epidemiological studies, resulting in potentially excessive additional cost.

- The level of chemical exposure must be documented as comprehensively and accurately as possible. In most instances, exposure should also be sufficiently high to be associated with potentially interpretable immunological changes, as most current clinical immunology assays have a limited sensitivity. Importantly, as the magnitude of environmental or occupational exposure is decreasing in many areas of the world, epidemiological immunotoxicity studies in this context might well result in falsely negative results.

- Sample acquisition at sites geographically distant from the investigator's laboratory is a major difficulty to be overcome when using available assays. The availability and/or implementation of assays is often more critical in the selection of biomarkers than scientific considerations.

- Finally, and largely based on the various points raised above, immunotoxicity studies are extremely expensive if all these criteria are dealt with adequately.

In fact, no epidemiological studies using immunological endpoints have so far been carried out, and the few published studies were limited to field studies. A few illustrative examples of such studies include the consequences of consumption of fish contaminated with organochlorine derivatives (Svensson *et al.*, 1994), the immunotoxicity of heavy metals in the Great Lakes (Bernier *et al.*, 1995), or the adverse influence of halogenated organic compounds in hobby fishermen (Løvik *et al.*, 1996). Nevertheless, the need for conducting properly designed epidemiological studies has been stressed and this is likely to be a major challenge for immunotoxicologists in the near future (Van Loveren *et al.*, 1997).

Use of immunological endpoints in clinical trials of medicinal products

Surprisingly, immunological endpoints have been very seldom included in the clinical trials of medicinal products, except when the product under scrutiny had expected or intended immunotherapeutic activity (Buehles, 1998). A logical consequence of performing routine non-clinical immunotoxicity evaluation would be to include specific clinical immunology endpoints in human subjects of clinical trials either to confirm the suggested lack of immunotoxicity, or to assess the reality and magnitude of observed immunological changes in animals.

Use of disease clusters and sentinel events

Disease clusters

Environmental epidemiological studies are typically designed to identify greater rates of disease in relation to certain exposures. In some instances, the distribution of diseases is not uniform in a given population and the question arises as to whether case occurrence within a certain location is increased over expected number: cases seemingly grouped within a certain location are called disease clusters (Aldrich *et al.*, 1993).

Possible immunotoxicity has so far not been evoked as a cause of disease cluster, but the finding that CD4[+] T lymphocytes were decreased in a small group of women residing in households with groundwater contaminated by the pesticide aldicarb (Fiore *et al.*, 1986), can be considered an example of an immunotoxic event cluster. In fact, numbers are usually very small so that it is very difficult, if ever possible to demonstrate increased occurrence rates. Cluster studies have essentially been devoted to certain types of cancers, such as leukaemias, and have often been a matter of excessive hype by the media and the public.

Sentinel events

Because of the limitations in available biomarkers of immunotoxicity, the surveillance of sentinel events was suggested to be another potentially useful approach to identify human immunotoxicants (National Research Council 1992; Sundaran and Wing 1995; Van Loveren *et al.*, 1997).

The concept of sentinel events has already been used in other areas of medicine, such as infectious diseases, Parkinson's disease, cerebrovascular accidents or suicide, as well as in the fields of occupational toxicology, veterinary toxicology, or mutagenicity (Descotes *et al.*, 1996a). A sentinel toxic event is defined as an adverse event with a sufficient understanding in the underlying pathophysiological mechanism(s) to indicate or, at least, strongly suggest that a chemically-mediated injury to a given target organ or function has occurred, and that a causal relationship between the toxic injury and a documented chemical exposure might exist. Relatively rare events should be preferably selected as sentinel events, to avoid excessive background noise. One major specificity of sentinel programmes is that they are based on spontaneous case reporting and therefore rely on less stringent and expensive procedures than epidemiological studies. Positive as well as negative consequences can however be delineated: a wider fraction of the general population can be attained using a sentinel approach, but it is not possible to calculate incidence rates, even though changes in reported sentinel event rates may be suggestive of a causality trend, despite many possible biases. It is also easier for general practitioners and clinicans with many unrelated medical activities to take part in sentinel programmes than in large-scale epidemiological studies, but sentinel programmes, whatever the efforts being paid, can never be as comprehensive and reliable as disease registries or epidemiological studies.

Sentinel event studies are potentially valuable tools for the immunotoxicologist because immune-mediated diseases have been reported to be associated with drug treatments and/or exposure to occupational and environmental chemicals. An example of such a programme was started in France (Pham *et al.*, 1995; Descotes *et al.*, 1996a): the organ-specific autoimmune diseases, myasthenia, pemphigus and thyroiditis, and the systemic autoimmune diseases, lupus, Gougerot–Sjögren syndrome, rheumatoid arthritis, Sharp's syndrome, dermatopolymyositis and scleroderma were selected. Collaborating physicians were provided with specific reporting forms to obtain detailed information on the patient's medical history, past drug and chemical exposures, and clinical and biological diagnostic criteria of the disease. When a sufficient amount of information is judged to be available, the decision is made whether the new case fulfils the predetermined criteria. As no procedure was purposedly included to determine a causal relationship, no conclusion can be expected to be drawn on this aspect. When the number of certain sentinel event/documented exposure cases exceeds a calculated limit based on the random distribution of similar events in the dataset, a warning signal is generated. Epidemiological studies and/or experimental studies are then warranted to confirm the validity of the warning signal.

Surveillance of exposed populations

The surveillance of selected groups of the population exposed to recognised immunotoxicants is another major area of interest and investigation for human immunotoxicology. When the immune system has been shown to be a primary target organ of toxicity, it could be useful to search for immunological changes indicative of excessive chemical exposure with a greater risk of developing adverse health consequences. However, only a limited number of toxicants, such as tributyltin oxide which induced thymic atrophy in rats, were shown to be immunotoxic at dose levels inducing no or limited toxicity on other target organs (Penninks and Pieters, 1996).

Chronic exposure to beryllium was shown to induce an immune-mediated lung disease clinically and biologically close to sarcoidosis (Newman, 1998). Despite some discrepancies, the beryllium lymphocyte transformation test was shown to be positive in 3 to 4 per cent of exposed workers. For various reasons, such as genetic susceptibility, not all exposed individuals develop a clinical lung disease. Therefore, the beryllium lymphocyte transformation test can be a useful tool to predict individuals who are more likely to develop the disease among exposed workers and who are the most likely to require immediate cessation of exposure.

Validation of non-clinical results

Although it might seem rather surprising to recommend the use of human data to validate the results of non-clinical immunotoxicity studies, there is a need for such a validation. In most other areas of toxicology, human findings used to be obtained prior to experimental studies so that animal models were subsequently designed to investigate the toxic phenomenon in depth. For instance, a number of carcinogenic chemicals were identified following the development of cancers in occupationally exposed workers, whereas the teratogenic effects of thalidomide proved instrumental to the further development of predictive animal models in reproduction toxicology. In the field of immunotoxicology, the vast majority of data has been obtained in non-clinical studies so that the relevance of these findings for immunotoxicity risk assessment remains debatable, despite major advances in this field.

By gaining a more comprehensive clinical experience in the immunotoxicity of xenobiotics, reference immunotoxicants can be identified and prove instrumental for the comparison of experimental results. This is in agreement with the last version of the immunotoxicity testing guidelines for pesticides under the EPA's Toxic Substances Contact Act (*Federal Register*, 1997), requiring that cyclophosphamide is used as a positive control. It is however unsure that the use of positive controls as potent as cyclophosphamide (or cyclosporin) is always appropriate and might not result in inadequate conclusions.

Biomarkers of human immunotoxicity

Biomarkers are essential tools which enable toxicologists to measure the exposure to hazardous chemicals and the magnitude of toxic responses, and/or to predict the likelihood of these responses (Timbrell, 1998). Biomarkers can thus be divided into three possibly overlapping categories: biomarkers of exposure; biomarkers of toxicity; and biomarkers of susceptibility. Biomarkers may be simple or complex, specific to mammals or other species, and used in *in vitro* or *in vivo* systems.

During the past two decades, extensive efforts have been devoted by clinical immunologists to design assays helpful for the laboratory diagnosis of immune diseases in patients with primary or secondary immune deficiencies, or a history of hypersensitivity reaction or autoimmune disease (Brostoff *et al.*, 1991; Colvin *et al.*, 1995; Gooi and Chapel, 1990; Graziano and Lemanske, 1989; Janeway and Travers, 1994). Initially, methods used in clinical immunology have been directly adapted to human or clinical immunotoxicology so that many immune endpoints or biomarkers have been proposed for use in investigating whether drug treatments or occupational and environmental exposures can be associated with clinically significant adverse effects on the human immune system. Because so many endpoints have been proposed and because most of these endpoints have initially been developed to address clinical immunology issues specifically, a selection of relevant endpoints was needed to provide a cost-effective list of biomarkers to be included in field or epidemiological studies. Among panels of experts which were convened to select immune endpoints as biomarkers of immunotoxicity, the conclusions of two panels deserve particular attention.

Immune endpoints adopted by the Agency for Toxic Substances and Disease Registry

The panel of experts convened by the Agency for Toxic Substances and Disease Registry (ATSDR) recommended that available immune tests are divided into three categories or levels (Straight *et al.*, 1994): basic tests (level 1), focused tests (level 2) and research tests (level 3) (see Table 19.1).

Basic tests were selected because they can easily be performed and are relevant to the detection and diagnosis of immune deficiency, hypersensitivity and autoimmunity. They include serum levels of antinuclear antibodies, C-reactive protein, IgG, IgM and IgA, total proteins, total white blood cell count, total lymphocyte count, total eosinophil count, and $CD4^+$ and $CD8^+$ lymphocyte counts.

Focused tests were selected for the follow-up of abnormal basic test results. They are intended to be used in specific patients or groups of patients. Recommended tests focusing on immune deficiency include measurement of serum antibody levels to a given antigen (such as a vaccine) and serum isohaemagglutinins, complement CH_{50} assay, granulocyte tetrazolium dye reduction assay, mitogen-induced lymphocyte proliferation assay and skin tests. Recommended tests focusing on hypersensitivity include total and specific IgE serum levels, beryllium lymphocyte transformation test, leukocyte histamine release assay and skin tests. Finally, recommended tests focusing on autoimmunity include serum levels of antithyroglobulin, antimitochondrial, antiphospholipid, and anti-smooth muscle antibodies and rheumatoid factor. All other tests are considered as research tests (level 3) and are not recommended for use in epidemiological immunotoxicity studies.

Recommendations of the US National Research Council

The Subcommittee of the US National Research Council reviewed the status of biomarkers of immunotoxicity (National Research Council, 1992). Recommendations were that 'clinical studies in humans are needed to determine the relationship between chemical exposures and immune-mediated diseases'. Emphasis was put on the use of sensitive and validated assays and standardised case definitions. The need for methods to extrapolate animal data to human beings more accurately was stressed, as was the need for better quantification of exposure, and improved identification of confounding and risk factors.

That the recommendations of this panel of experts focused more on research needs than on possible modalities of use of biomarkers of immunotoxicity is not surprising, as

Table 19.1 Immune endpoints adopted by the ATSDR (adapted from Straight *et al.*, 1994)

Basic immune test battery (level 1)

Serum
   Antinuclear antibodies
   C-reactive protein
   Immunoglobulins (IgG, IgM, IgA)
   Total proteins

Blood
   Total white blood cell count
   Total lymphocyte count
   $CD4^+$ and $CD8^+$ T lymphocytes

Focused tests of immune function (level 2)

Immune deficiency
   Antibody response to a given antigen (e.g. vaccine)
   Complement CH_{50} assay
   Serum isohaemagglutinins
   Granulocyte nitroblue tetrazolium dye reduction assay
   Mitogen-induced lymphocyte proliferation assay
   Skin testing (delayed reading)

Hypersensitivity
   Total and specific serum IgE levels
   Beryllium lymphocyte transformation test
   Leukocyte histamine release assay
   Skin testing (immediate reading)

Autoimmunity
   Antithyroglobulin antibodies
   Antimitochondrial antibodies
   Antiphospholipid antibodies
   Anti-smooth muscle antibodies
   Rheumatoid factor

---

available biomarkers have repeatedly been claimed to be poorly relevant and helpful, if at all (Descotes *et al.*, 1995, 1996b; Kimber, 1995; Vos and Van Loveren, 1995).

References

ALDRICH, T., DRANE, W. and GRIFFITH, J. (1993) Disease clusters. In: *Environmental Epidemiology and Risk Assessment* (Aldrich, T., Griffith, J. and Cooke, C., eds), pp. 61–82. New York: Van Rostrand-Reinhold.

BERNIER, J., BROUSSEAU, P., KRZYSTYNIAK, K., TRYPHONAS, H. and FOURNIER, M. (1995) Immunotoxicity of heavy metals in relation to Great Lakes. *Environ. Health Perspect.*, **103**, Suppl 9, 23–34.

BIAGINI, R.E. (1998) Epidemiology studies in immunotoxicity evaluations. *Toxicology*, **129**, 37–59.

BIAGINI, R.E., WARD, E.M., VOGT, R. and HENNINGSEN, G.M. (1994) Targeted epidemiology and clinical assessment studies of the immune system in humans. In: *Immunotoxicology and Immunopharmacology*, 2nd edition (Dean, J.H., Luster, M.I., Munson, A.E. and Kimber, I., eds), pp. 31–50. New York: Raven Press.

BROSTOFF, J., SCALDING, G.K., MALE, D. and ROITT, I.M. (1991) *Clinical Immunology.* London: Mosby.

BUEHLES, W.C. (1998) Application of immunologic methods in clinical trials. *Toxicology*, **129**, 73–89.

BURREL, R., FLAHERTY, D.K. and SAUERS, L.J. (1992) *Toxicology of the Immune System. A Human Approach.* New York: Van Rostrand Reinhold.

COLVIN, R.B., BAHN, A.K. and McCLUSKEY, R.T. (1995) *Diagnostic Immunology*, 2nd edition. New York: Raven Press.

DESCOTES, J., NICOLAS, B., PHAM, E. and VIAL, T. (1995) Assessment of immunotoxic effects in humans. *Clin.Chem.*, **41**, 1870–1873.

DESCOTES, J., NICOLAS, B., PHAM, E. and VIAL, T. (1996a) Sentinel screening in human immunotoxicology. *Arch. Toxicol.*, **Suppl. 18**, 29–34.

DESCOTES, J., NICOLAS, B., VIAL, T. and NICOLAS, J.F. (1996b) Biomarkers of immunotoxicity in man. *Biomarkers*, **1**, 77–80.

Federal Register (1997) Toxic Substances Control Act Test Guidelines. Final Rule. **62**, 43819–43864.

FIORE, M.C., ANDERSON, H.A., HONG, R., GOLUBJATNIKOV, R., SEISER, J.E., NORSTROM, D., *et al.* (1986) Chronic exposure to aldicarb-contaminated groundwater and human immune function. *Environ. Res.*, **41**, 633–645.

GOOI, H.G. and CHAPEL, H. (1990) *Clinical Immunology: A Practical Approach.* Oxford: Oxford University Press.

GRAZIANO, F.M. and LEMANSKE, R.F. (1989) *Clinical Immunology.* Baltimore: Williams and Wilkins.

GRIFFITH, J. and ALDRICH, T. (1993) Epidemiology: the environmental influence. In: *Environmental Epidemiology and Risk Assessment* (Aldrich, T., Griffith, J. and Cooke, C., eds), pp. 13–26. New York: Van Rostrand-Reinhold.

GRIFFITH, J., ALDRICH, T. and DUNCAN, R.C. (1993) Epidemiologic research methods. In: *Environmental Epidemiology and Risk Assessment* (Aldrich, T., Griffith, J. and Cooke, C., eds), pp. 27–60. New York: Van Rostrand-Reinhold.

HERNBERG, S. (1992) *Introduction to Occupational Epidemiology.* Chelsea, MI: Lewis Publishers.

KIMBER, I. (1995) Biomarkers of immunotoxicity in man. *Hum. Exp. Toxicol.*, **14**, 148–149.

JANEWAY, C.A. and TRAVERS, P. (1994) *Immunobiology: the Immune System in Health and Disease.* London: Current Biology.

LØVIK, M., JOHANSEN, H.R., GAARDER, P.I., BECHER, G., AABERGE, I.S., GDYNIA, W. and ALEXANDER, J. (1996) Halogenated organic compounds and the human immune system: preliminary report on a study in hobby fishermen. *Arch. Toxicol.*, **Suppl. 18**, 15–20.

NATIONAL RESEARCH COUNCIL (1992) *Biologic Markers in Immunotoxicology.* Washington DC: National Academy Press.

NEWCOMBE, D.S., ROSE, N.R. and BLOOM, J.C. (1992) *Clinical Immunotoxicology.* New York: Raven Press.

NEWMAN, L.S. (1998) Beryllium. In: *Immunotoxicology of Environmental and Occupational Metals* (Zelikoff, J.T. and Thomas, P.T., eds), pp. 27–40. London: Taylor & Francis.

PENNINKS, A.H. and PIETERS, R.H.H. (1996) Immunotoxicity of organotins. In: *Experimental Immunotoxicology* (Smialowicz, R.J. and Holsapple, M.P., eds), pp. 229–244. Boca Raton: CRC Press.

PHAM, E., NICOLAS, B., VIAL, T. and DESCOTES, J. (1995) Modeling of warning procedures to detect the toxic effects of drugs and chemicals in man: application to a sentinel disease program in immunotoxicology. *Toxicol. Model.*, **1**, 207–218.

SALVAGGIO, J.E. (1990) The impact of allergy and immunology on our expanding industrial environment. *J. Allergy Clin. Immunol.*, **85**, 689–699.

STRAIGHT, J.M., KIPEN, H.M., VOGT, R.F. and AMLER, R.W. (1994) *Immune Function Test Batteries for Use in Environmental Health Field Studies.* Atlanta: Agency for Toxic Substances and Diseases Registry.

SUNDARAN, S. and WING, M. (1995) Conclusions of an international workshop on environmental toxicology and human health. *Hum. Exp. Toxicol.*, **14**, 85–86.

SVENSSON, B.G., HALLBERG, T., NILSSON, A., SCHÜTZ, A. and HAGMAR, L. (1994) Parameters of immunological competence in subjects with high consumption of fish contaminated with persistent organochlorine compounds. *Int. Arch. Occup. Environ. Health*, **65**, 351–358.

TAUBES, G. (1995) Epidemiology faces its limits. *Science*, **269**, 164–169.

TIMBRELL, J.A. (1998) Biomarkers in toxicology. *Toxicology*, in press.

VAN LOVEREN, H., SRÁM, R. and NOLAN, C. (1997) *Environment and Immunity*. Air Pollution Epidemiology Reports Series, Report 11. Brussels: European Commission.

20

Immunotoxicology and Wildlife

Recently, the potential immunotoxicity of environmental chemicals in wildlife became a matter of growing concern. Because fish are exposed to pollutants present in contaminated water, they can be used as sentinel animals or biomarkers of immunotoxicity. The epidemic which killed thousand of seals from the North Sea was another important impetus to immunotoxicity studies in wildlife. In contrast, limited information is available on the immunotoxicity of veterinary medicinal drugs and environmental chemicals on domestic animals (Black and McVey, 1992; Koller, 1979). The possible impact of antibiotics on specific and non-specific immune responses was the reason for most studies. Overall, the reported effects in domestic animals are similar to those noted in humans.

Fish immunotoxicology

Because of the development of aquaculture, concern arose regarding the economic impact of diseases in fish, and this in turn served as a basis for the development of research in fish immunology and microbiology (Anderson and Zeeman, 1995; Van Muiswinkel *et al.*, 1985; Zelikoff, 1994). In addition, the impact of water contamination by chemical pollutants on fish health status is an expanding research area.

Fish immune system and diseases

The immune system of fish globally performs the same functions as in mammals, particularly the prevention of microbial diseases and the destruction of neoplastic cells for maintained good health. Because fish are very diverse (over 20 000 known species), there are many differences in their immune system. In various aspects, the immune system of fish is however comparable to that of mammals (Van Muiswinkel *et al.*, 1985; Zapata and Cooper, 1990). The fish immune system is formed of non-specific and specific immune effector mechanisms, the latter including humoral and cellular immune responses. Thus, fish can produce antigen-specific antibodies, mount a delayed-type hypersensitivity response, or reject allografts. Their lymphocytes can proliferate in response to mitogens and release cytokines, while their granulocytes and macrophages can phagocytose foreign pathogens. No lymph nodes or bone marrow are found in fish, but an equivalent of the haematopoietic tissue is located in several areas of the body, such as the spleen, thymus, head and kidney. Fish have circulating white blood cells, which are functionally and morphologically similar to those of mammals.

The most common diseases in fish related to pollution are skin diseases, such as lymphocystis, papillomas, fin rot and skin ulcers (Van Muiswinkel *et al.*, 1985; Vos *et al.*, 1988). These diseases can be easily identified and because they are caused by bacteria or viruses, the role of the immune system is obvious. Liver tumours have also been suggested to correlate with water pollution. Nevertheless, no conclusive correlations between these diseases and environmental pollution have so far been obtained. Although laboratory studies are helpful to identify the immunotoxic effects of pesticides, heavy metals, and chlorinated hydrocarbons present, for instance, in harbour sediment, sewage sludge, or pulp mill effluent, it is far more difficult to delineate the role of immunotoxic chemicals in field conditions, as other factors, such as migratory patterns, temperature, and stress, may interfere with the development of the disease.

Immunotoxicity studies in fish

Most screening and functional immunotoxicological techniques routinely used in mammals have been successfully applied to the study of fish immunology and immunotoxicology (Zelikoff, 1994, 1998). However, these techniques are generally under development in research laboratories and are therefore not yet applicable to field studies. Immunosuppression is the primary, if not only, focus of immunotoxicity studies in fish. Exposure to aquatic environmental pollutants has been suggested or shown to be associated with many diseases and toxic disorders in fish. Such deleterious exposures can have a major economical impact with respect to aquaculture, but adverse effects on fish health status can also be indicative of possible health consequences on human health, and therefore a current impetus for fish immunotoxicology studies is the identification of fish biomarkers as sentinel effects of immunotoxicity for mammals, including man (Wester *et al.*, 1994; Zelikoff, 1998).

Although the value of biomarkers of immunotoxicity in fish is still largely unclear, because their role in the aetiology of fish diseases is not well understood, but also because the value of biomarkers of immunotoxicity in general has so far not been established, a number of immune parameters have been proposed and used for immunotoxicity evaluation in fish.

Selection of species

A wide range of species can be used and the selection is primarily based on the investigator's experience and the laboratory's background. The preferred species are trout, carp, flounder and dab, because they are more easily manageable due to their size, which enables blood and tissue sampling. Smaller species, e.g. guppies, have the advantage of easy husbandry and lowered cost, but blood and tissue sampling are difficult.

Histopathology

White blood cell counts and lymphocyte subset analysis can easily be performed in a blood sample drawn from a live fish. In fish of adequate size, the spleen can be easily removed and weighed, but the lymphoid tissue is poorly developed in the spleen of most fish species. Thymus weight is seldom used in fish immunotoxicology because of its complex localisation in most fish species. Thymus morphology could be more useful, and thymus atrophy could be evidenced in several, but not all fish species, after exposure to tributyltin oxide, a prototypic immunotoxicant in rats (Zelikoff, 1993).

Melanomacrophage centres are widely distributed throughout the body of fish. They are formed of clusters of macrophages, the function of which is not yet fully understood. They have been suggested to be primitive analogues of lymphoid follicles, and tend to increase with age and after stress.

Non-specific immunity

Macrophage function tests, such as chemiluminescence, chemotaxis and phagocytosis, are considered to be useful indicators of immunotoxicity. Fish macrophages indeed share most morphological and functional features of mammalian macrophages, and a number of studies evidenced the adverse impact of environmental chemicals on the various functions of fish macrophages.

Humoral immunity

Humoral immunity can be assessed by the determination of total circulating immunoglobulin levels or the specific antibody response to a standard antigen, such as sheep erythrocytes. Agglutination and ELISA are the most commonly used techniques. The plaque-forming cell is less commonly used. LPS can induce the proliferation of fish B lymphocytes.

Cellular immunity

Cellular immunity is quite similar in fish to cellular immunity in mammals. Evaluation of cellular immunity can be best performed using mitogen-induced lymphocyte proliferation assays, but *in vivo* models such as foreign tissue allograft rejection, have also been used, although rarely.

Immunotoxicity of chemicals in fish

Fish are exposed to an enormous variety of chemicals due to discharges to rivers and lakes, marine dumpings and atmospheric fallout. The impact of aquatic chemical pollutants on aquaculture in certain areas, such as the Mississippi river, has been emphasised. Only the main fish immunotoxicants will be considered as reviews are available elsewhere (Anderson and Zeeman, 1995; Zeeman and Brindley, 1981; Zelikoff, 1994).

Immunotoxicity of metals in fish

The toxic effects of metals on fish have been investigated extensively (Zelikoff, 1993), but limited information is available regarding the influence of metals on the immune responsiveness of fish. Lead and cadmium have been shown to be markedly immuno-suppressive, with reduced antibody response as seen in mammals, even though a few experimental data showed immunoenhancing properties as in both laboratory animals and human beings. The 'biphasic' effects of heavy metals, depending on the level of exposure and the time of exposure with respect to antigen injection, previously discussed in this volume, were also seen to occur in fish. Impaired cellular immunity was reported following cadmium exposure, but inconsistently. Cadmium was also shown to impair non-specific immune responses, such as trout macrophage phagocytosis.

Manganese, a trace element necessary for life in mammals and fish, was shown to be highly toxic in fish, but nevertheless to increase consistently NK cell activity, lymphocyte proliferation and macrophage functions. The immunotoxic effects of nickel have seldom been investigated in fish, but nickel has been proved to suppress humoral immunity and macrophage functions.

Immunotoxicity of pesticides in fish

Pesticides are widely used and water contamination by pesticides is widespread. The organochlorine insecticides are major pollutants of aquatic ecosystems because of their persistence. No immunotoxic effect of lindane was reported, in contrast to DDT, which was shown to decrease antibody response and more importantly to be associated with an increased incidence of parasitic and fungal diseases in fish. Several organophosphate insecticides, such as methylparathion, malathion and trichlorphon, were shown to depress both humoral and cellular immunity. The carbamate insecticide carbaryl was shown to suppress both humoral and cellular immune responses, whereas aldicarb, a suspected but unconfirmed immunotoxicant in mammals, did not produce marked immunotoxic effects in fish. The herbicide atrazine was shown to induce no significant immunotoxic effects in fish. In contrast, the fungicide pentachlorphenol proved to be potently immunotoxic.

Immunotoxicity of halogenated aromatic hydrocarbons in fish

Contamination of rivers, lakes and seas by halogenated aromatic hydrocarbons is also widespread. In contrast to mammals, fish do not seem to be very sensitive to the immunotoxic effects of biphenyls on humoral immunity. However, dramatic changes in lymphoid organs and tissues have been evidenced in mammals as well as fish. Susceptibility of fish to infectious diseases was also shown to be impaired following biphenyl exposure.

Immunotoxicity of veterinary and human medicinal products in fish

Few medicinal products have been tested for their effects on fish immune responses. Overall, data obtained with antimicrobials in fish mimic those already obtained in laboratory animals or man, particularly in that the relevance of findings with respect to host resistance towards microbial infections is unknown. Corticosteroids were shown to be immunosuppressive in fish, resulting in thymic involution and increased susceptibility to infections. Alkylating agents proved to be immunosuppressive in trout. The beneficial or supposedly beneficial role of vitamins and antioxidants on immune responsiveness was not more conclusively evidenced in fish than in mammals.

Sea mammals

Seals from seas of the northern Europe suffer from various diseases in relation to environmental contamination of water (Olsson *et al.*, 1994). In 1988, an epizootic resulted in the deaths of approximately 20 000 harbour seals and several hundred grey seals in

Europe (Harwood, 1989). Affected animals presented with fever, cutaneous lesions, gastrointestinal dysfunction, neurological disorders and respiratory distress. The causative agent was identified to be a previously unidentified morbillivirus, named phocid distemper virus-1 or PDV-1. However, the contributing role of polyhalogenated aromatic hydrocarbons, including polychlorinated biphenyls, polychlorinated dibenzo-*p*-dioxins and polychlorinated dibenzofurans, was evidenced by comparing results of immune function tests (NK cell activity, T lymphocyte function, delayed-type hypersensitivity and antibody response) in seals fed herring from either the relatively uncontaminated Atlantic Ocean or the highly contaminated Baltic Sea (Ross *et al.*, 1996). These results indicate that current levels of persistent lipophilic contaminants present an immunotoxic risk to harbour seals inhabiting many coastal areas in Europe and presumably North America as well.

Similarly, dolphins from the Atlantic Ocean or the Mediterranean Sea have been shown to be exposed to high levels of chemical contaminants in water. Although studies on the immune status of dolphins in relation to environmental contamination have seemingly not been performed, the immunotoxic effects of contaminants are likely to play a role in the epizootic which recently killed many dolphins from the Mediterranean Sea (Borrell *et al.*, 1996).

Other sea mammals whose immune functions, such as NK cell activity, are under investigation, include beluga whales (De Guise *et al.*, 1998).

Perspectives

Toxic chemicals have clearly been shown to induce a wide variety of diseases, both in animals and human beings. Immunological mechanisms are likely to be involved in some of these diseases, while the causal relationship of other diseases with immune changes is still speculative. Nevertheless, data gained from the assessment of immunotoxic chemicals in wildlife species is further evidence that immunotoxicity is potentially associated with significant adverse health effects. Fish and sea mammals however are not the only wildlife species exposed to immunotoxic environmental pollutants. Birds are also exposed to a variety of environmental chemicals. Although bird immunotoxicology is not an expanding area of research as is fish immunotoxicology, the impact of chemicals on the immune competence of birds and the resulting consequence on the outcome of acute exposure to petrochemicals and oil-related toxicants is a matter of growing concern (Briggs *et al.*, 1996; Rocke *et al.*, 1984).

Because wildlife species are living within our contaminated world, they can also serve as sentinel species, either to identify environmental contamination resulting in overt immunotoxicity, or to determine and control the level of contamination better with no associated immunotoxic consequences for the health of human beings (as well as that of domestic animals and wildlife species). An obvious disadvantage is the limited value of current biomarkers of immunotoxicity, but as soon as more reliable biomarkers are available, immunotoxicity studies in wildlife species are likely to become a very important aspect of both immunotoxicology and ecotoxicology.

References

ANDERSON, D.P. and ZEEMAN, M.G. (1995) Immunotoxicology in fish. In: *Fundamentals of Aquatic Toxicology: Effects, Environmental Fate, and Risk Assessment.* (Rand, G., ed.), pp. 371–404. Washington DC: Taylor & Francis.

BLACK, R.D. and MCVEY, D.S. (1992) Immunotoxicity in the bovine animal: a review. *Vet. Hum. Toxicol.*, **34**, 438–442.

BORRELL, A., AGUILAR, A., CORSOLINI, S. and FOCARDI, S. (1996) Evaluation of toxicity and sex-related variation of PCB levels in Mediterranean striped dolphins affected by an epizootic. *Chemosphere*, **32**, 2359–2369.

BRIGGS, K.T., YOSHIDA, S.H. and GERSCHWIN, M.E. (1996) The influence of petrochemicals and stress on the immune system of seabirds. *Regul. Toxicol. Pharmacol.*, **23**, 145–155.

DE GUISE, S., ROSS, P.S., OSTERHAUS, A.D.M.E., MARTINEAU, D., BELAND, P. and FOURNIER, M. (1998) Immune functions in beluga whales (Delphinapterus leucas): evaluation of natural killer cell activity. *Vet. Immunol. Immunopathol.*, **58**, 345–354.

HARWOOD, J. (1989) Lessons from the seal epidemics. *New Scientist*, 18 February, 38–42.

KOLLER, L.D. (1979) Effects of environmental contaminants on the immune system. *Adv. Vet. Sci. Comp. Med.*, **23**, 267–295.

OLSSON, M., KARLSSON, B. and AHNLAND, E. (1994) Diseases and environmental contaminants in seals from the Baltic and the Swedish west coast. *Sci. Total Environ.*, **154**, 217–227.

ROCKE, T.E., YUILL, T.M. and HINSDILL, R.D. (1984) Oild and related toxicant effects on mallard immune defenses. *Environ. Res.*, **33**, 343–352.

ROSS, P., DE SWART, R., ADDISON, R., VAN LOVEREN, H., VOS, J. and OSTERHAUS, A. (1996) Contaminant-induced immunotoxicity in harbour seal: wildlife at risk? *Toxicology*, **112**, 157–169.

VAN MUISWINKEL, W.B., ANDERSON, D.P., LAMERS, C.H.J., EGBERTS, E., VAN LOON, J.J.A. and IJSSEL, J.P. (1985) Fish immunology and fish health. In: *Fish Immunology* (Manning, M.J. and Tatner, M.F., eds), pp. 316–333. London: Academic Press.

VOS, J.G., VAN LOVEREN, H., WESTER, P.W. and VETHAAK, A.D. (1988) The effects of environmental pollutants on the immune system. *Eur. Environ. Rev.*, **2**, 2–7.

WESTER, P.W., VETHAAK, A.D. and VAN MUISWINKEL, W.B. (1994) Fish as biomarkers in immunotoxicology. *Toxicology*, **86**, 213–232.

ZAPATA, E.L. and COOPER, E.L. (1990) *The Immune System: Comparative Histophysiology*. New York: John Wiley & Sons.

ZEEMAN, M.G. and BRINDLEY, W.A. (1981) Effects of toxic agents upon fish immune system: a review. In: *Immunologic Considerations in Toxicology*, vol. II (Sharma, R.P., ed.), pp. 1–60. Boca Raton: CRC Press.

ZELIKOFF, J.T. (1993) Metal pollution-induced immunomodulation in fish. *Annu. Rev. Fish Dis.*, **2**, 305–325.

ZELIKOFF, J.T. (1994) Fish immunotoxicology. In: *Immunotoxicology and Immunopharmacology*, 2nd edition (Dean, J.H., Luster, M.I., Munson, A.E. and Kimber, I., eds), pp. 71–95. New York: Raven Press.

ZELIKOFF, J.T. (1998) Biomarkers of immunotoxicity in dish and other non-mammalian sentinel species: predictive value for mammals? *Toxicology*, **129**, 63–71.

Index